T0203098

SpringerBriefs in Applied Sciences and Technology

Computational Intelligence

Series editor

Janusz Kacprzyk, Warsaw, Poland

About this Series

The series "Studies in Computational Intelligence" (SCI) publishes new developments and advances in the various areas of computational intelligence—quickly and with a high quality. The intent is to cover the theory, applications, and design methods of computational intelligence, as embedded in the fields of engineering, computer science, physics and life sciences, as well as the methodologies behind them. The series contains monographs, lecture notes and edited volumes in computational intelligence spanning the areas of neural networks, connectionist systems, genetic algorithms, evolutionary computation, artificial intelligence, cellular automata, self-organizing systems, soft computing, fuzzy systems, and hybrid intelligent systems. Of particular value to both the contributors and the readership are the short publication timeframe and the world-wide distribution, which enable both wide and rapid dissemination of research output.

More information about this series at http://www.springer.com/series/10618

Neha Yadav · Anupam Yadav
Manoj Kumar

An Introduction to Neural Network Methods for Differential Equations

Springer

Neha Yadav
Department of Applied Sciences
ITM University
Gurgaon, Haryana
India

Anupam Yadav
Department of Sciences and Humanities
National Institute of Technology Uttarakhand
Srinagar, Uttarakhand
India

Manoj Kumar
Department of Mathematics
Motilal Nehru National Institute of Technology
Allahabad
India

ISSN 2191-530X ISSN 2191-5318 (electronic)
SpringerBriefs in Applied Sciences and Technology
ISBN 978-94-017-9815-0 ISBN 978-94-017-9816-7 (eBook)
DOI 10.1007/978-94-017-9816-7

Library of Congress Control Number: 2015932071

Springer Dordrecht Heidelberg New York London

Printed on acid-free paper

Springer Science+Business Media B.V. Dordrecht is part of Springer Science+Business Media (www.springer.com)

Preface

Artificial neural networks, or neural networks, represent a technology that is rooted in many disciplines like mathematics, physics, statistics, computer science and engineering. Neural networks have various applications in the area of mathematical modelling, pattern recognition, signal processing and time-series analysis, etc. It is an emerging field for researchers and scientists in the industry and academics to work on. Also, many problems in science and engineering can be modelled with the use of differential equations such as problems in physics, chemistry, biology and mathematics. Due to the importance of differential equations, many methods have been developed in the literature for solving them, but they have their own shortcomings.

This book introduces a variety of neural network methods for solving differential equations arising in science and engineering. Emphasis is placed on the deep understanding of the neural network techniques, which have been presented in a mostly heuristic and intuitive manner. This approach will enable the reader to understand the working, efficiency and shortcomings of each neural network technique for solving differential equations.

The objective of this book is to provide the readers with a sound understanding of the foundations of neural network, comprehensive introduction to neural network methods for solving differential equations along with the recent developments in the techniques. The main purpose to write this textbook is stated in its title *An Introduction to Neural Network Methods for Differential Equations*. This book aims to get started with the neural network techniques for solving differential equations easily, quickly and pleasantly to beginners, regardless of their background—physics, chemistry, mathematics or engineering. This book is a comprehensive text on neural network methods for solving differential equations, and the subject matter is presented in an organized and systematic way. The book may serve as a background for readers who do not have in-depth knowledge of differential equations and neural networks together with building a basic skill set that can be used to be master in it. Our presentation in the book is aimed at developing the insights and techniques that are most useful for attacking new problems. To compile this book, we had to borrow ideas from different sources and the credit goes to all the original developers of these networks; we have presented a list of references for each section.

This book has been compiled in four chapters. The Introduction provides a glimpse of the organization of the book and a general introduction. Chapter 1 consists of a brief overview of differential equations and the physical problems arising in science and engineering. Chapter 2 illustrates the history of neural networks starting from the 1940s beginning to the 1980s renewed enthusiasm. A general introduction to neural networks and learning technologies is presented in Chap. 3. This chapter also includes a description of multilayer perceptron and its learning methods. In Chap. 4, we introduce the different neural network methods for solving differential equations. The recent developments in all the techniques is also presented in this section. The conclusion is also presented at the end of Chap. 4, which concludes the topics presented in the book. An exhaustive list of references is given at the end of the book.

Neha Yadav
Anupam Yadav
Manoj Kumar

Contents

Introduction

A series of problems in many scientific fields can be modelled with the use of differential equations such as problems in physics, chemistry, biology, economics, etc. Although model equations based on established physical laws may be constructed using analytical tools and are frequently inadequate for the purpose of obtaining their closed form solution. Due to the importance of differential equations many methods have been proposed in the existing literature for their solution. Principal numerical methods available for solving differential equations are Finite difference method (FDM), Finite element method (FEM), Finite volume method (FVM), the boundary element method (BEM), etc. These methods generally require discretisation of the domain into a number of finite elements (FEs), which is not a straightforward task.

In contrast, for FE-type approximation, neural networks can be considered as approximation schemes where the input data for a design of network consist of only a set of unstructured discrete data points. Thus an application of neural network for solving differential equations can be regarded as a mesh-free numerical method. The solution via neural network is differentiable, closed analytic form and easily used in any subsequent calculation. Most other techniques offer a discrete solution or a solution of limited differentiability. This book presents the general introduction to neural networks and a brief description of different neural network methods for solving ordinary and partial differential equations.

Neural networks are simplified models of the biological nervous system and therefore have drawn their motivation from the kind of computing performed by a human brain. In general, the neural network is a highly interconnected network of a large number of processing elements called neurons in an architecture inspired by the brain. The neural network learns by examples and thus can be trained to acquire knowledge about the system. Once the training has been performed appropriately, the network can be put to effective use for solving 'unknown' instances of the problem. Neural networks adopt various learning mechanisms among which supervised and unsupervised learning methods have turned out to be very popular.

Neural networks have been successfully applied to problems in the fields of pattern recognition, image processing, forecasting and optimization, etc.

Initially, most of the work in solving differential equations using neural network is restricted to the case of solving the system of algebraic equations which result from the discretisation of the domain. The solution of a linear system of equations is mapped onto the architecture of a neural network and the solution to the system of given equations is then obtained by the minimization of the network's energy function. Another approach to the solution of differential equations is based on the fact that certain types of splines, for instance B_1 splines, can be derived by the superposition of piecewise linear activation functions. The solution of differential equations using B_1 splines as basis functions can be obtained by solving a system of linear or nonlinear equations in order to determine the coefficients of splines. Such a solution is mapped directly on the architecture of a feedforward neural network by replacing each spline with the sum of piecewise linear activation functions that correspond to the hidden units. This method considers local basis function and in general requires many splines in order to yield accurate solution. Furthermore, it is not easy to extend these techniques to multidimensional domains.

In this book we present different neural network methods for solution of differential equations, which provides many attractive features towards the solution: (i) The solution using neural network is differentiable, closed analytic form and easily used in any subsequent calculation; (ii) Method is general and can be applied to solve ordinary as well as partial differential equations with higher order complexities; (iii) Method requires less number of model parameters than any other technique and hence requires less memory space; (iv) Provides a solution with very good generalization properties.

The objective of this book is to provide the reader with a sound understanding of the foundations of neural network and a comprehensive introduction to different neural network methods for solving differential equations. Our presentation is aimed at developing the insights and techniques that are most useful for attacking new problems. However, the matter presented in this book is available in different books and research articles but we summarized the important useful material in an effective manner, which can serve as an introduction to new researchers and be helpful both as a learning tool and as a reference.

The structure of the book is as follows. The book is divided into four chapters.

Chapter 1, entitled "Overview of Differential Equations", introduces fundamentals of differential equation problems with some appropriate examples. This chapter also explains some existing numerical methods with examples for the solution of differential equations.

Chapter 2, entitled "History of Neural Networks", presents the origin of Neural Network in the existing literature.

Chapter 3, entitled "Preliminaries of Neural Networks", introduces the fundamentals of neural networks along with their learning algorithms and major architectures.

Chapter 4, entitled "Neural Network Methods for Solving Differential Equations", contains different neural network methods for solving differential equations of various kinds and complexities. This chapter also contains some worked out numerical examples arising in real-life applications.

MATLAB code for the solution of differential equations based on neural network has been also given in the Appendix section.

Chapter 1
Overview of Differential Equations

Abstract This chapter presents a general introduction to differential equations together with its boundary conditions. In general, a differential equation is an equation which involves the derivatives of an unknown function represented by a dependent variable. It expresses the relationship involving the rates of change of continuously changing quantities modeled by functions and are used whenever a rate of change (derivative) is known. A brief introduction to different numerical methods in the existing literature like finite difference, finite element, shooting method and spline based method is also presented.

Keywords Ordinary differential equation · Partial differential equation · Dirichlet boundary condition · Neumann boundary condition · Mixed boundary condition

The term "differential equations" (aequatio differentialis) was initiated by Leibnitz in 1676. A solution to a differential equation is a function whose derivatives satisfy equation [1–6].

1.1 Classification of Differential Equations

The differential equations can be categorized in ordinary differential equation (ODE), partial differential equation (PDE), delay differential equation (DDE), stochastic differential equation (SDE) and differential algebraic equation (DAE) which are defined as follows:

1.1.1 Ordinary Differential Equations

An ordinary differential equation (ODE) is a differential equation in which the unknown function is a function of a single independent variable. It implicitly describes a function depending on a single variable and the ODE expresses a

N. Yadav et al., *An Introduction to Neural Network Methods for Differential Equations*, SpringerBriefs in Computational Intelligence,
DOI 10.1007/978-94-017-9816-7_1

relation between the solution and one or more of its derivatives. Beside the ODE, usually one or more additional (initial) conditions are needed to determine the unknown function uniquely. The most general form of an ordinary differential equation of nth order is given by

$$\frac{d^n y}{dx^n} = f\left(x, y, \frac{dy}{dx}, \frac{d^2y}{dx^2}, \ldots, \frac{d^{n-1}y}{dx^{n-1}}\right),$$

which is termed as ordinary because there is only one independent variable.

1.1.2 Partial Differential Equations

A partial differential equation (PDE) is a relation involving an unknown function of at least two independent variables and its partial derivatives with respect to those variables. Partial differential equations are used to formulate and solve problems that involve unknown functions of several variables, such as the propagation of sound or heat, electrostatics, electrodynamics, fluid flow, elasticity or more generally any process that is distributed in space or distributed in space and time. In general, A partial differential equation (PDE) is an equation involving functions and their partial derivatives.

1.1.3 Delay Differential Equations

A delay differential equation (DDE) is a special type of functional differential equation which is similar to ordinary differential equation but in delay differential equation derivative of the unknown function at a certain time is given in terms of the values of the function at previous times. The solution of delay differential equation therefore requires of knowledge of not only the current state, but also of the state at certain time previously.

1.1.4 Stochastic Differential Equations

A stochastic differential equation (SDE) is a differential equation in which one or more of the terms are a stochastic process, thus resulting in a solution which is itself a stochastic process. Stochastic differential equation used to model diverse phenomenon such as fluctuating stock prices or physical system subject to thermal fluctuations.

1.1.5 *Differential Algebraic Equations*

A differential algebraic equation (DAE) is a generalized form of ordinary differential equation which involves an unknown function and its derivatives. This type of equation arises in the mathematical modeling of wide variety of problem from engineering and science such as optimal control, chemical process control, incompressible fluids etc.

1.2 Types of Differential Equation Problems

1.2.1 *Initial Value Problem*

An initial value problem is one in which the dependent variable and its possible derivatives are specified initially or at the same value of independent variable in the equation. Initial value problems are generally time-dependent problems.

For example: If the independent variable is time over the domain [0, 1], an initial value problem would specify a value of $y(t)$ at time 0. Physically, in the middle of a still pond if somebody taps the water with a known force that would create a ripple and gives us an initial condition.

1.2.2 *Boundary Value Problem*

A boundary value problem is one in which the dependent variable and its possible derivatives are specified at the extreme of the independent variable. For steady state equilibrium problems, the auxiliary conditions consist of boundary conditions on the entire boundary of the closed solution domain.

For example: If the independent variable is time over the domain [0, 1], a boundary value problem would specify values for $y(t)$ at both $t = 0$ and $t = 1$. If the problem is dependent on both space and time, then instead of specifying the value of the problem at a given point for all time, the data could be given at a given time for all space. For example, the temperature of an iron bar with one end kept at absolute zero and the other end at freezing point of water would be a boundary value problem. There are three types of boundary conditions:

1.2.2.1 Dirichlet Boundary Condition

In Dirichlet boundary condition, values of the function are specified on the boundary. For example if an iron rod has one end held at absolute zero then the value of the problem would be known at that point in the space. A Dirichlet boundary condition imposed on an ordinary or a partial differential equation specifies the values of a solution is to take on the boundary of the domain and finding the solution the solution of such a equations are known as the Dirichlet problem.

For example: Let us consider a case of an partial differential equation:

$$\begin{aligned} &\frac{\partial^2 u}{\partial x^2}+\frac{\partial^2 u}{\partial y^2}=f(x,y) \quad \text{in } \Omega \\ &u(x,y)=A \qquad \text{on } \partial\,\Omega \end{aligned} \tag{1.1}$$

where, A is some number. The boundary condition given in Eq. (1.1) represents a Dirichlet boundary condition as the value of the function u(x,y) is specified on the boundary.

1.2.2.2 Neumann Boundary Condition

In Neumann boundary condition, values of the function are specified on the derivative normal to the boundary. For example if one iron rod had heater at one end then energy would be added at constant rate but the actual temperature would not be known. A Neumann boundary condition imposed on the ordinary or partial differential equation specifies the derivative values of solution are to take on the boundary of the domain.

For example:

$$\begin{aligned} &\frac{\partial^2 u}{\partial x^2}+\frac{\partial^2 u}{\partial y^2}=f(x,y) \quad \text{in } \Omega \\ &\frac{\partial u}{\partial n}=w \qquad \text{on } \partial\,\Omega \end{aligned} \tag{1.2}$$

1.2.2.3 Mixed Boundary Condition

Mixed boundary conditions are the linear combination of Dirichlet and Neumann boundary conditions and also known as the Cauchy boundary condition. A mixed boundary condition imposed on an ordinary or partial differential equation specifies both the values of a differential equation is to take on the boundary of the domain and the normal derivative at the boundary. It corresponds to imposing both Dirichlet and Neumann boundary condition:

$$\begin{aligned} &\frac{\partial^2 u}{\partial x^2}+\frac{\partial^2 u}{\partial y^2}=f(x,y) \quad \text{in } \Omega \\ &w_1\frac{\partial u}{\partial n}+w_2 u=w \qquad \text{on } \partial\,\Omega \end{aligned} \tag{1.3}$$

1.3 Differential Equations Associated with Physical Problems Arising in Engineering

As the world turns, things change, Mountains erode, river beds change, machines break down, the environment becomes more polluted, populations shift, economics fluctuate, technology advances. Hence any quantity expressible mathematically over a long time must change as a function of time. As a function of time, relatively speaking, there are many quantities which change rapidly, such as natural pulsation of a quartz crystal, heart beats, the swing of a pendulum, chemical explosions, etc.

When we get down to the business of quantitative analysis of any system, our experience shows that the rate of change of a physical or biological quantity relative to time has vital information about the system. It is this rate of change which plays a central role in the mathematical formulation of most of the physical and biological models amenable to analysis.

Engineering problems that are time-dependent are often described in terms of differential equations with conditions imposed at single point (initial value problems); while engineering problems that are position dependent are often described in terms of differential equations with conditions imposed at more than one point (boundary value problems). Some of the motivational examples encountering in many engineering fields are:

(i) Coupled L-R electric circuits,
(ii) Coupled systems of springs,
(iii) Motion of a particle under a variable force field,
(iv) Newton's second law in dynamics (mechanics),
(v) Radioactive decay in nuclear physics,
(vi) Newton's law of cooling in thermodynamics,
(vii) The wave equation,
(viii) Maxwell's equations in electromagnetism
(ix) The heat equation in thermodynamics,
(x) Laplace's equation, which defines harmonic functions,
(xi) The beam deflections equation,
(xii) The draining and coating flows equations etc.

1.4 General Introduction of Numerical Methods for Solving Differential Equations

In the field of mathematics the existence of solution in many cases is guaranteed by various theorems, but no numerical method for obtaining those solutions in explicit and closed form is known. In view of this the limitations of analytic methods in practical applications have led the evolution of numerical methods and there are various numerical methods for different type of complex problems which have no analytical solution.

Analytical solutions, when available, may be precise in themselves, but may be of unacceptable form because of the fact that they are not amenable to direct interpretation in numerical terms, in which case the numerical analyst may attempt to derive a method for effecting that interpretation in a satisfactory way. Numerical techniques to solve the boundary value problems include some of the following methods:-

1.4.1 Shooting Method

These are initial value problem methods. In this method, boundary value problems are transformed into two initial value problems by adding sufficient number of conditions at one end and adjust these conditions until the given conditions are satisfied at the other end. The solution of these two initial value problems is determined by such methods as the Taylor series, Runge-Kutta etc. and the required solution of the given boundary value problem is given by the addition of the two solutions obtained by solving initial value problems.

For example: Let us consider a boundary value problem given as:

$$y'' = f(t, y, y') \quad \text{with} \quad y(a) = \alpha \quad \text{and} \quad y(b) = \beta \tag{1.4}$$

We can solve this problem by taking the related initial value problem with a guess as to the appropriate initial value $y'(a)$ and integrate the equation to obtain an approximate solution hoping that $y(b) = \beta$. If $y(b) \neq \beta$ then the guesses value of $y'(a)$ can be change by trying again. This process is called shooting and there are different ways for doing it systematically. If we consider the guessed value of $y'(a)$ is k, so the corresponding boundary value problem becomes

$$y'' = f(t, y, y') \quad \text{with} \quad y(a) = \alpha \quad \text{and} \quad y'(\text{a}) = k \tag{1.5}$$

The solution of this initial value problem will be denoted by y_k and our objective is to select k such that $y_k(b) = \beta$. Let us consider $\phi(k) = x_k(b) - \beta$, so that our objective is to simply solve the equation $\phi(k) = 0$ for k which can be solve by any of the method for solving non linear equations e.g. Bisection method, Secant method etc. Each value of $\phi(k)$ is computed by numerically solving an initial value problem.

1.4.2 Finite Difference Method

In finite difference method (FDM), functions are represented by their values at certain grid points and derivatives are approximated through differences in these values. For the finite difference method, the domain under consideration is represented by a finite subset of points. These points are called "nodal points" of the grid. This grid is almost always arranged in (uniform or non-uniform) rectangular manner. The differential equation is replaced by a set of difference equations which are solved by direct or iterative methods.

For example: consider the second order boundary value problem

$$y'' = f(x, y, y') \tag{1.6}$$

with

$$\begin{aligned} y(a) &= \alpha \quad \text{or} \quad y'(a) = \alpha \\ y(b) &= \beta \quad \text{or} \quad y'(b) = \beta \end{aligned} \tag{1.7}$$

Approximating the derivatives at the mesh point by finite differences gives:

$$\frac{y_{i-1} - 2y_i + y_{i+1}}{h^2} = f\left(x_i, y_i, \frac{y_{i+1} - y_i}{2h}\right), \quad i = 1, 2, \ldots, n \tag{1.8}$$

with

$$y_1 = \alpha \text{ or } \frac{y_2 - y_0}{2h} = \alpha \tag{1.9}$$

$$y_n = \beta \text{ or } \frac{y_{n+1} - y_{n-1}}{2h} = \beta \tag{1.10}$$

Rewriting Eq. (1.8) by elimination the points outside the domain as

$$y_0 - 2y_1 + y_2 - h^2 f\left(x_1, y_1, \frac{y_2 - y_0}{2h}\right) = 0 \tag{1.11}$$

$$y_{i-1} - 2y_i + y_{i+1} - h^2 f\left(x_i, y_i, \frac{y_{i+1} - y_{i-1}}{2h}\right) = 0, \quad i = 2, 3, \ldots, n-1 \tag{1.12}$$

$$y_{n-1} - 2y_n + y_{n+1} - h^2 f\left(x_n, y_n, \frac{y_{n+1} - y_{n-1}}{2h}\right) = 0 \tag{1.13}$$

The boundary conditions on y are replaced by $y_1 - \alpha = 0$ and $y_n - \beta = 0$, y_0 and y_{n+1} are obtained from Eqs. (1.9) and (1.10) and then substituted in Eqs. (1.11) and (1.12) respectively. Thus we can obtain the set of n simultaneous algebraic equations with n unknowns which can be solved by any of the method applicable for solving set of algebraic equations.

1.4.3 Finite Element Method

The finite element method is a numerical method like finite difference method but it is more general and powerful for the real world problems that involve complicated boundary conditions. In finite element method (FEM), functions are represented in terms of basis functions and the differential equations are solved in its integral (weak) form. In the finite element method the domain under consideration is partitioned in a finite set of elements $\{\Omega_i\}$ so that $\{\Omega_i \cap \Omega_j\} = \phi$ for $i \neq j$, and $\cup\overline{\Omega}_i = \overline{\Omega}$. Then the function is approximated by piecewise polynomial of low degree. Further they are constructed so that their support extends only over a small number of elements. The main reason behind taking approximate solution on a collection of sub domains is that it is easier to represent a complicated function as a collection of simple polynomials. To illustrate the Finite element method let us consider the following boundary value problem:

$$u'' = u + f(x), \quad x<0<1 \tag{1.14}$$

with

$$u(0) = 0 \quad \text{and} \quad u(1) = 0 \tag{1.15}$$

Finite element methods finds piecewise polynomial approximation $v(x)$ to the solution of Eq. (1.14) which can be represented by the equation

$$v(x) = \sum_{j=1}^{m} \alpha_j \phi_j(x) \tag{1.16}$$

where $\phi_j(x), j = 1, 2, \ldots, m$ are specified functions that are piecewise continuously differentiable called basis functions and α_j are unknown constants. In case of Galerkin method Eq. (1.14) is multiplied by ϕ_i, $i = 1, 2, \ldots, m$ and integrate the resulting equation over the domain [0, 1]

$$\int_0^1 [u''(x) - u(x) - f(x)]\phi_i(x)dx = 0, \quad i = 1, 2, \ldots, m \tag{1.17}$$

Since the functions $\phi_i(x)$ satisfies the boundary conditions Eq. (1.17) becomes:

$$\int_0^1 u'(x)\phi_i'(x)dx + \int_0^1 [y(x) + f(x)]\phi_i(x)dx = 0, \quad i = 1, 2, \ldots, m \tag{1.18}$$

For any two function we define

$$(\eta, \psi) = \int_0^1 \eta(x)\psi(x)\mathrm{dx} \tag{1.19}$$

Using Eqs. (1.19) and (1.18) becomes

$$(u', \phi_i') + (u, \phi_i) + (f, \phi_i) = 0, \quad i = 1, 2, \ldots, m \tag{1.20}$$

Equation (1.19) is called the weak form of Eq. (1.13). If $v(x)$ is given by Eq. (1.16) then (1.20) becomes

$$\left(\sum_{j=1}^{m} a_j\phi_j', \phi_i'\right) + \left(\sum_{j=1}^{m} a_j\phi_j, \phi_i\right) + (f, \phi_i) = 0 \tag{1.21}$$

Solution of Eq. (1.21) gives the vector a, which specifies Galerkin approximation.

1.4.4 Finite Volume Method

The finite volume method is used to represent and evaluate differential equations in the form of algebraic equations. Finite Volume refers to the small volume surrounding each node point on a mesh. In the case of Finite volume method values are calculated at the discrete places of meshed geometry as in the Finite difference method. In this method, volume integrals in a differential equation that contain a divergence term are converted to surface integrals, using the divergence theorem. These terms are evaluated as fluxes at the surfaces of each finite volume and since the flux entering in a given volume is identical to that leaving the adjacent volume, the method is conservative in nature. Finite volume method has an advantage over the finite difference method that it does not require a structured mesh and also the boundary conditions can be applied non-invasively. This method is powerful on non uniform grids and in calculations where the mesh moves to track interfaces.

1.4.5 Spline Based Method

Usually a spline is a piece-wise polynomial function defined in a region, such that there exist a decomposition of the region into the sub regions in each of which the function is a polynomial of some degree. In spline based methods, the differential equation is discretized by using approximate methods based on spline. The end conditions are derived for the definition of spline. The algorithm developed not only approximates the solutions, but their higher order derivatives as well.

For example: Consider the two point boundary value problem of the form:

$$-\frac{d}{dx}\left[p(x)\frac{du}{dx}\right] = g(x) \qquad (1.22)$$
$$u(a) = u(b) = 0$$

where $p \in C^1[a,b], p > 0$ and $g \in C[a,b]$. To solve the Eq. (1.22) with spline method, we consider a uniform mesh Δ with nodal points x_i with equal intervals. Consider a non polynomial function $s_\Delta(x)$ for each segment $[x_i, x_{i+1}]$, $i = 0, 1, \ldots, N-1$ of the following form:

$$s_\Delta(x) = a_i + b_i(x - x_i) + c_i \sin\tau(x - x_i) + d_i \cos\tau(x - x_i), \quad i = 0, 1, \ldots, N \qquad (1.23)$$

where a_i, b_i, c_i and d_i are constants and τ is free parameter. Let u_i be an approximation to $u(x_i)$ which can be obtained by the segment $s_\Delta(x)$ of the mixed splines function passing through the points (x_i, u_i) and (x_{i+1}, u_{i+1}). To derive the expressions for the coefficient of Eq. (1.23) in terms of u_i, u_{i+1}, M_i and M_{i+1}, we first denote:

$$s_\Delta(x_i) = u_i, \quad s_\Delta(x_{i+1}) = u_{i+1}, \quad s''_\Delta(x_i) = M_i, \quad s''_\Delta(x_{i+1}) = M_{i+1} \qquad (1.24)$$

Thus from algebraic manipulation we get the following equation:

$$a_i = u_i + \frac{M}{\tau^2}, \; b_i = \frac{u_{i+1} - u_i}{h} + \frac{M_{i+1} - M_i}{\tau\theta}$$
$$c_i = \frac{M_i \cos\,\theta - M_{i+1}}{\tau^2 \sin\,\theta}, \; d_i = -\frac{M_i}{\tau^2} \qquad (1.25)$$

where $\theta = \tau h$ and $i = 0, 1, \ldots, N-1$.

Using the continuity of first derivative we get the following equation:

$$\alpha M_{i+1} + 2\beta M_i + \alpha M_{i-1} = \frac{1}{h^2}(u_{i+1} - 2u_i + u_{i-1}) \qquad (1.26)$$

where,

$$\alpha = \frac{1}{h^2}(\theta \operatorname{cosec}\,\theta - 1), \; \beta = \frac{1}{h^2}(1 - \theta\cot\theta)$$

Hence by using the moment of spline in Eq. (1.19) we obtain

$$M_i + q_i u_i' = f_i$$

By approximating the first derivative of u and substituting the equations into the Eq. (1.26) we get the tri-diagonal system of equation, which can be solve by any of the method for solving system of equations.

1.4.6 Neural Network Method

Neural network methods can solve both ordinary and partial differential equations that relies on the function approximation capabilities of feed forward neural networks and results in a solution written in a closed analytic form. This form employs a feed forward neural network as a basic approximation element whose parameters are adjusted to minimize an appropriate error function. Training of the neural network can be done by any optimization technique which in turn requires the computation of the gradient of the error with respect to the network parameters. In this method, a trial solution of the differential equation is written as a sum of two parts. The first part satisfies the initial or boundary conditions and contains no adjustable parameters. Second part is constructed so as to not affect the initial or boundary conditions and involves a feed forward neural network by containing adjustable parameters. Hence by construction the trial solution, the initial or boundary conditions are satisfied and the network is trained to satisfy the differential equation.

1.5 Advantages of Neural Network Method for Solving Differential Equations

A neural network based model for the solution of differential equations provides the following advantages over the standard numerical methods:

(a) The neural network based solution of a differential equation is differentiable and is in closed analytic form that can be used in any subsequent calculation. On the other hand most other techniques offer a discrete solution or a solution of limited differentiability.
(b) The neural network based method to solve a differential equation provides a solution with very good generalization properties.
(c) Computational complexity does not increase quickly in the neural network method when the number of sampling points is increased while in the other standard numerical methods computational complexity increases rapidly as we increase the number of sampling points in the interval.

(d) The method is general and can be applied to the systems defined on either orthogonal box boundaries or on irregular arbitrary shaped boundaries.
(e) Model based on neural network offers an opportunity to tackle in real time difficult differential equation problems arising in many sciences and engineering applications.
(f) The method can be implemented on parallel architectures.

Chapter 2
History of Neural Networks

Abstract Here we are presenting a brief history of neural networks, given in Haykin (Neural networks: a comprehensive foundation, 2002) [7], Zurada (Introduction to artificial neural systems, 2001) [8], Nielsen (Neurocomputing, 1990 [9] in terms of the development of architectures and algorithms that are widely used today. The history of neural networks has been divided in four stages: Beginning of neural networks, First golden age, Quiet Years and Renewed enthusiasm which shows the interplay among biological experimentation, modeling and computer simulation, hardware implementation.

Keywords Perceptron · ADALINE · Signal processing · Pattern recognition · Biological modeling · Neurocomputing

2.1 The 1940s: The Beginning of Neural Networks

The beginning of Neurocomputing is often taken to be the research article of McCulloch and Pitts [10] published in 1943, which showed that even simple types of neural networks could, in principle, compute any arithmetic or logical function, was widely read and had great influence. Other researchers, principally Norbert Wiener and von Neumann, wrote a book and research paper [11, 12] in which the suggestion was made that the research into the design of brain-like or brain-inspired computers might be interesting.

In 1949 Hebb wrote a book [13] entitled *The Organization of Behaviour* which pursued the idea that classical psychological conditioning is ubiquitous in animals because it is a property of individual neurons. This idea was not itself new, but Hebb took it further than anyone before him had by proposing a specific learning law for the synapses of neurons. Hebb then used this learning law to build a qualitative explanation of some experimental results from psychology. Although there were many other people examining the issues surrounding the neurocomputing in the 1940s and early 1950s, their work had more the effect of setting the

N. Yadav et al., *An Introduction to Neural Network Methods for Differential Equations*, SpringerBriefs in Computational Intelligence,
DOI 10.1007/978-94-017-9816-7_2

stage for later developments than of actually causing those developments. Typical of this era was the construction of first neurocomputer (the *Snark*) by Marvin Minsky in 1951. The Snark did operated successfully from a technical stand point but it never actually carried out any particularly interesting information processing functions.

2.2 The 1950s and 1960s: The First Golden Age of Neural Networks

The first successful neuro-computer (the Mark I perceptron) was developed during 1957 and 1958 by Frank Rosenblatt, Charles Wightman, and others. As we know it today, Rosenblatt as the founder of Neurocomputing. His primary interest was pattern recognition. Besides inventing the perceptron, Rosenblatt also wrote an early book on Neurocomputing, *Principles of Neurodynamics* [14].

Slightly later than Rosenblatt, but cut from similar cloth, was Bernard Widrow. Widrow, working with his graduate students (most notably Marcian E. "Ted" Hoff, who later went on to invent the microprocessor) developed a different type of neural network processing element called ADALINE, which was equipped with a powerful new learning law which, unlike the perceptron leaning law, is still in widespread use. Widrow and his students applied the ADALINE successfully to a large number of toy problems, and produced several films of their successes. Besides Rosenblatt and Widrow, there were a number of other people during the late 1950s and early 1960s who had substantial success in the development of neural network architectures and implementation concepts.

Notwithstanding the considerable success of these early Neurocomputing researchers, the field suffered from two glaringly obvious problems. *First*, the majority of researchers approached the subject from a qualitative and experimental point of view. This experimental emphasis resulted in a significant lack of rigor and a looseness of thought that bothered many established scientists and engineers who established the field. *Second*, an unfortunate large fraction of neural networks researchers were carried away by their enthusiasm in their statements and their writings. For example, there were widely publicized predictions that artificial brains were just a few years away from development, and other incredible statements.

Besides the hype and general lack of rigor, by the mid 1960s researchers had run out of good ideas. The final episode of this era was a campaign led by Marvin Minsky and Seymour Papert to discredit neural network research and divert neural network research funding to the field of "Artificial Intelligence". The campaign was waged by the means of personal persuasion by Minsky and Papert and their allies, as well as by limited circulation of unpublished technical manuscript (which was further published in 1969 by Minsky and Papert as the book *Perceptrons* [15]).

The implicit thesis of *Perceptrons* was that essentially all neural networks suffer from the same "fatal flaw" as the perceptron; namely the inability to usefully compute certain essentials predicates such as XOR. To make this point the authors reviewed several proposed improvements to the perceptron and showed that these were also unable to perform well. They left the impression that neural network research had been proven to be a dead end.

2.3 The 1970s: The Quiet Years

In spite of Minsky and Papert's demonstration of the limitations of perceptrons, research on neural network continued. A great deal of neural network research went on under the headings of adaptive signal processing, pattern recognition, and biological modeling. In fact, Many of the current leaders in the field began to publish their work during 1970s. Examples include Amari [16], Fukushima [17], Grossberg [18] and Klopf and Gose [19]. These people, and those who came in over the next 13 years, were the people who put the field of neural network on a firm footing and prepared the way for the renaissance of the field.

2.4 The 1980s: Renewed Enthusiasm

By the early 1980s many Neurocomputing researchers became bold enough to begin submitting proposals to explore the development of neuro-computers and of neural network applications. In the years 1983–1986 John Hopfield, an established physicist of worldwide reputation, had become interested in neural networks a few years earlier. Hopfield wrote two highly readable papers on neural networks in 1982 [20] and 1984 [21] and these, together with his many lectures all over the world, persuaded hundreds of highly qualified scientists, mathematicians, and technologists to join the emerging field of neural networks.

In 1986, with the publication of the "PDP books" (Parallel Distributed Processing, Volumes I and II, edited by Rumelhart and McClelland [22]), the field exploded. In 1987, the first open conference on neural networks in modern times, the IEEE International Conference on Neural Networks was held in San Diego, and the International Neural Network Society (INNS) was formed. In 1988 the INNS journal *Neural Networks* was founded, followed by Neural Computation in 1989 and the IEEE Transactions on Neural Networks in 1990.

Chapter 3
Preliminaries of Neural Networks

Abstract In this chapter brief introduction to neural network has been given along with some basic terminologies. We explain the mathematical model of neural network in terms of activation functions. Different architectures of neural network like feed forward, feed backward, radial basis function network, multilayer perceptron neural network and cellular network etc., is described. Backpropagation and other training algorithms have been also discussed in this chapter.

Keywords Neural network · Feed forward · Recurrent network · Particle swarm optimization · Genetic algorithm · Backpropagation algorithm

3.1 What Is Neural Network?

A neural network is a parallel distributed information processing structure in the form of a directed graph, (directed graph is a geometrical object consisting of a set of points called nodes along with asset of directed line segments called links between them) with the following sub-definitions and restrictions:

(i) The nodes of the graphs are called processing elements.
(ii) The links of the graphs are called connections. Each connection functions as instantaneous unidirectional signal-conduction path.
(iii) Each processing element can receive any number of incoming connections.
(iv) Each processing element can have any number of outgoing connections, but the signals in all of these must be the same. In effect, each processing element has a single output connection that can branch out or fan out into copies to form multiple output connections, each of which carries the same identical signal.
(v) Processing elements can have local memory.
(vi) Each processing element possess a transfer function which can use local memory, can use input signals, and which produces the processing element's output signal.
(vii) Input signals to a neural network from outside the network arrive via connections that originate in the outside world.

N. Yadav et al., *An Introduction to Neural Network Methods for Differential Equations*, SpringerBriefs in Computational Intelligence,
DOI 10.1007/978-94-017-9816-7_3

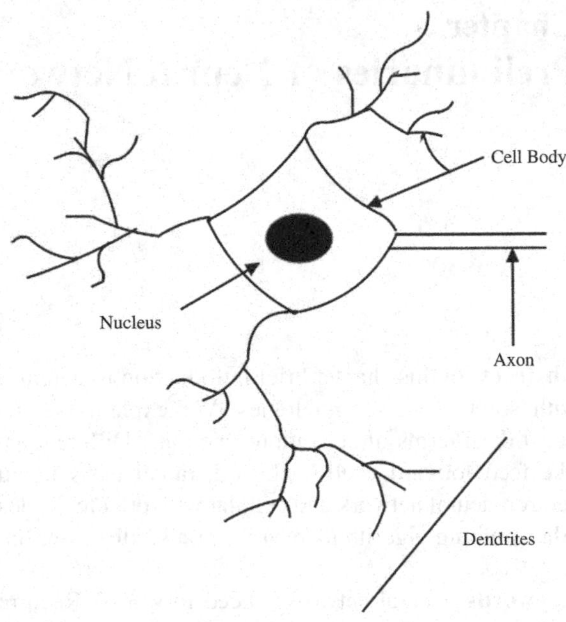

Fig. 3.1 Biological neural network

3.2 Biological Neural Network

Artificial neural networks emerged after the introduction of simplified neurons by McCulloch and Pitts in 1943. These neurons were presented as models of biological neurons and as conceptual components for circuits that could perform computational tasks. The basic model of the neuron is founded upon the functionality of a biological neuron. "Neurons are the basic signaling units of the nervous system" and "each neuron is a discrete cell whose several processes arise from its cell body".

The neuron has four main regions to its structure. The cell body, or soma, has two offshoots from it, the dendrites, and the axon, which end in presynaptic terminals. The cell body is the heart of the cell, containing the nucleus and maintaining protein synthesis. A neuron may have many dendrites, which branch out in a treelike structure, and receive signals from other neurons. A neuron usually only has one axon which grows out from a part of the cell body called the axon hillock. The axon conducts electric signals generated at the axon hillock down its length. These electric signals are called action potentials. The other end of the axon may split into several branches, which end in a presynaptic terminal. Action potentials are the electric signals that neurons use to convey information to the brain. All these signals are identical. Therefore, the brain determines what type of information is being received based on the path that the signal took. The brain analyzes the patterns of signals being sent and from that information it can interpret the type of information being received.

Myelin is the fatty tissue that surrounds and insulates the axon. At these nodes, the signal traveling down the axon is regenerated. This ensures that the signal traveling down the axon travels fast and remains constant. The synapse is the area of contact between two neurons. The neurons do not actually physically touch. They are separated by the synaptic cleft, and electric signals are sent through chemical interaction. The neuron sending the signal is called the presynaptic cell and the neuron receiving the signal is called the postsynaptic cell. Neurons can be classified by their number of processes (or appendages), or by their function [23]. Figure 3.1 represents the structure and functioning of biological neural network.

3.3 Artificial Neural Network

An artificial neural network (ANN) is an information–processing system that has certain performance characteristics in common with biological neural networks. Artificial neural networks have been developed as generalizations of mathematical models of human cognition or neural biology, based on the assumptions that:

(i) Information processing occurs at many simple connections called neurons.
(ii) Signals are passed between neurons over connection links.
(iii) Each connection link has an associated weight, which in a typical neural net, multiplies the signal transmitted.
(iv) Each neuron applies an activation function to its net input to determine its output signal.

The basic component of an artificial neural network is artificial neuron like biological neuron in biological neural network. A biological neuron may be modeled artificially to perform computation and then the model is termed as artificial neuron.

A neuron is the basic processor or processing element in a neural network. Each neuron receives one or more input over these connections (i.e., synapses) and produces only one output. Also this output is related to: the state of the neuron and its activation function. This output may fan out to several other neurons in the network. The inputs are the outputs i.e. activations of the incoming neurons multiplied by the connection weights or synaptic weights. Each weight is associated with an input of a network. The activation of a neuron is computed by applying a threshold function (popularly known as activation function) to the weighted sum of the inputs plus a bias. Figure 3.2 represents an artificial neuron.

3.4 Mathematical Model of Artificial Neural Network

A neuron N_i accepts a set of n inputs, $S = \{x_j | j = 1, 2, \ldots, n\}$. In Fig. 3.3, each input is weighted before reaching the main body of a neuron N_i by connection strength or weight factor w_{ij} for $j = 1, 2, \ldots, n$. In addition, it has a bias term w_o, a

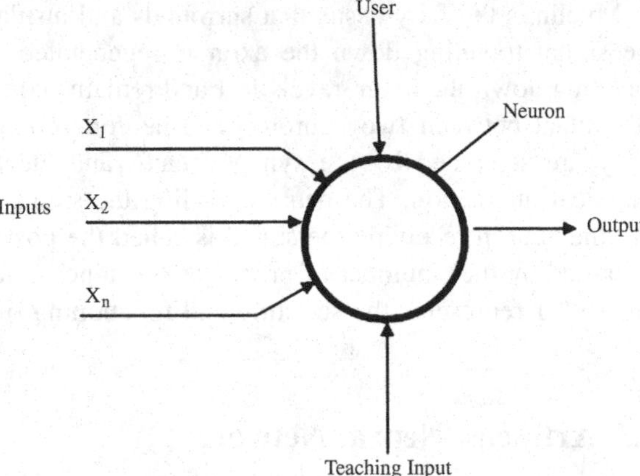

Fig. 3.2 An artificial neuron

threshold value θ_k, which has to be reached or exceeded for the neuron to produce an output signal. A function $f(s)$ acts on the produced weighted signal. This function is called the *activation function*. Mathematically, the output of the i-th neuron N_i is

$$O_i = f\left[w_o + \sum_{j=1}^{n} w_{ij}x_j\right] \tag{3.1}$$

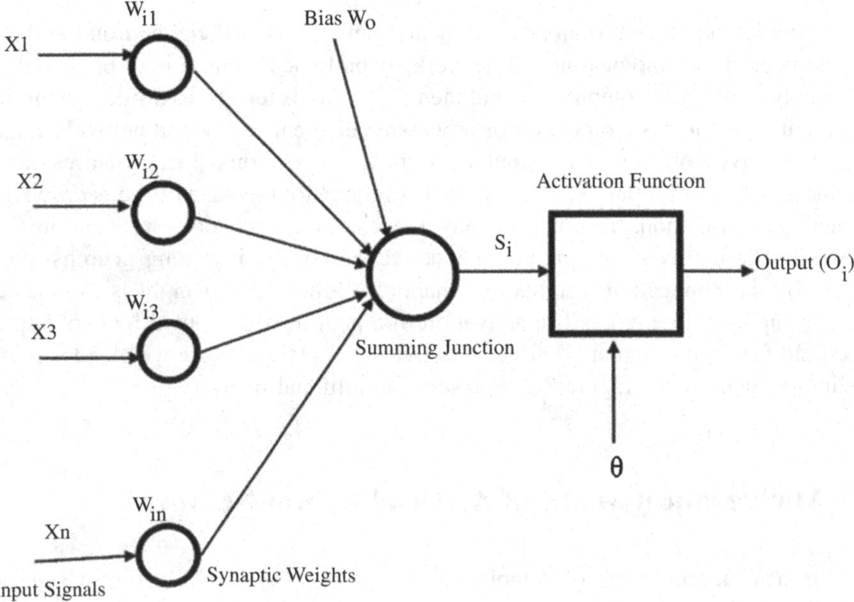

Fig. 3.3 Mathematical model of artificial neural network

And the neuron's firing condition is,

$$w_o + \sum_{j=1}^{n} w_{ij}x_j \geq \theta \tag{3.2}$$

Figure 3.3 shows detailed computational steps of the working principle of an artificial neuron in a neural network. Now the input signal for the i-th neuron N_i is

$$s_i = w_o + \sum_{j=1}^{n} w_{ij}x_j \tag{3.3}$$

This is obtained by adder function and the output signal obtained by activation function is:

$$O_i = f(s_i - \theta_i) \tag{3.4}$$

3.5 Activation Function

As we have discussed in the above section, the output signal is a function of the various inputs x_j and the weights w_{ij} which are applied to the neuron. Originally the neuron output function proposed as threshold function, however linear, sign, sigmoid and step functions are widely used output. Generally, inputs, weights, thresholds and neuron output could be real value or binary or bipolar. All inputs are multiplied to their weights and added together to form the net input to the neuron called *net*. Mathematically, we can write

$$\text{net} = w_{i1}x_1 + w_{i2}x_2 + \cdots w_{ij}x_j + \theta \tag{3.5}$$

where θ is a threshold value that is added to the neurons. The neuron behaves as activation or mapping function $f(net)$ to produce an output y which can be expressed as:

$$y = f(\text{net}) = f\left(\sum_{j=1}^{n} w_{ij}x_j + \theta\right) \tag{3.6}$$

where f is called the neuron activation function or the neuron transfer function. Some examples of the neuron activation functions are explained in Fig. 3.4.

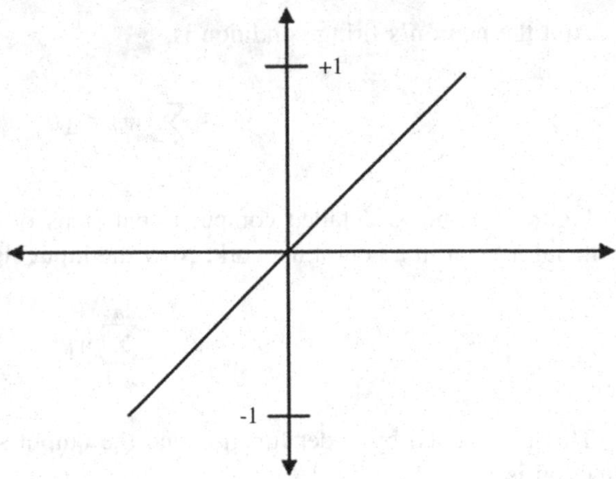

Fig. 3.4 Linear activation function

3.5.1 Linear Activation Function

The linear neuron transfer function is called the linear activation function or the ramp function, which is shown in Fig. 3.4.

$$y = f(\text{net}) = f\left(\sum_{j=1}^{n} w_{ij}x_j + \theta\right) = \text{net} \tag{3.7}$$

3.5.2 Sign Activation Function

Neuron transfer function is called sign activation function if the output is hard limited to the values +1 and −1, or sometimes 0 depending upon the sign of *net* as shown in Fig. 3.5. In this case the expression of the output y can be written as:

$$y = \begin{cases} 1 & \text{if net} \geq 0 \\ -1 & \text{if net} < 0 \end{cases} \tag{3.8}$$

3.5.3 Sigmoid Activation Function

It is an S shaped nonlinear smooth function, where input is mapped into values between +1 and 0. The neuron transfer function is shown in Fig. 3.6 and defined as:

$$y = \frac{1}{1 + e^{-Tx}} \tag{3.9}$$

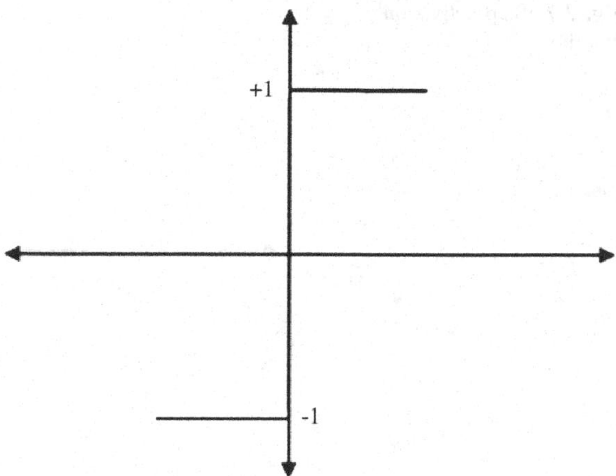

Fig. 3.5 Sign activation function

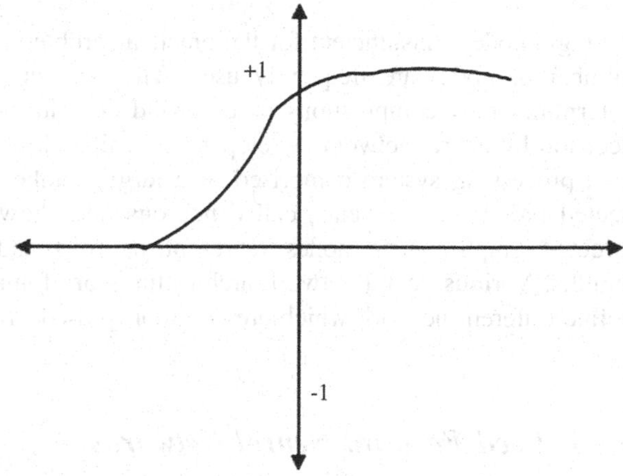

Fig. 3.6 Sigmoid activation function

3.5.4 Step Activation Function

In this case, the net neuron input is mapped into values between +1 and 0. The step activation function is shown in Fig. 3.7 and defined by:

$$y = \begin{cases} 1 & \text{if net} \geq 0 \\ 0 & \text{if net} < 0 \end{cases} \tag{3.10}$$

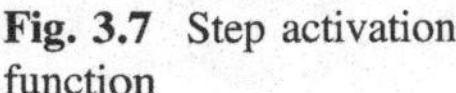

Fig. 3.7 Step activation function

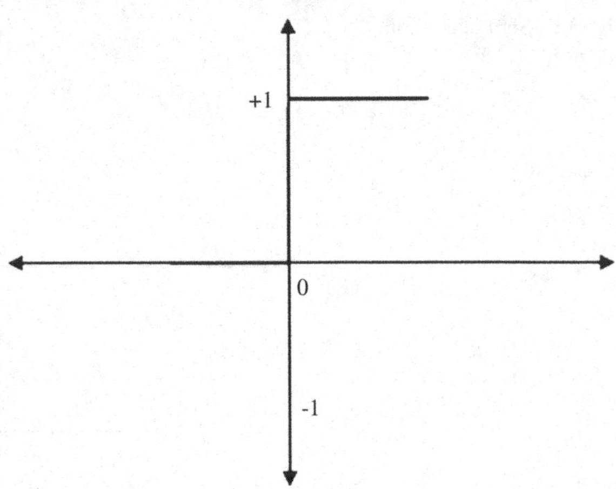

3.6 Neural Network Architecture

A single node is insufficient for the practical problems, and networks with a large number of nodes are frequently used. The way in which nodes are connected determines how computations proceed and constitutes an important early design decision by neural network developer. An artificial neural network is essentially a data processing system comprised of a large number of simple highly interconnected processing elements, called neurons and shown in Fig. 3.8–3.14 using a directed graph, where nodes represent neurons and edges represent synaptic lengths. Various neural network architectures are found in the literature. Here we define different network which are commonly used in current literature [24–26].

3.6.1 Feed Forward Neural Networks

Feed forward neural networks are the simplest form of artificial neural network. The feed forward neural network was the first and arguably simplest type of artificial network devised. In this network, the information moves in only one direction, forward, from the input nodes, through the hidden nodes (if any) and to the output nodes as shown in Fig. 3.8. There are no cycles or loops in the network. In a feed-forward system, processing elements (PE) are arranged into distinct layers with each layer receiving input from the previous layer and outputting to the next layer. Weights of direct feedback paths, from a neuron to itself are zero. Weights from a neuron to a neuron in a previous layer are also zero. Weights for the forward paths may also be zero depending on the specific network architecture, but they do not need to be in every case. Mathematically, we can express that,

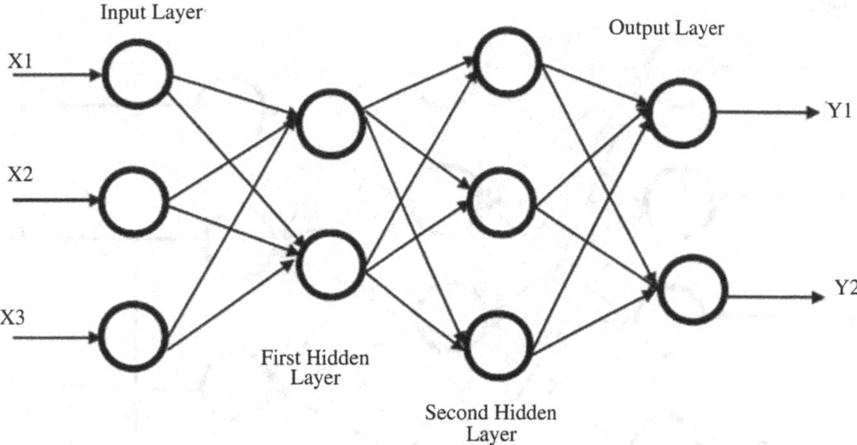

Fig. 3.8 A feed-forward network

$$\begin{aligned} w_{ij} &= 0 \quad \text{if} \quad i = j \\ w_{ij} &= 0 \quad \text{if} \quad \text{layer } i \leq \text{layer } j \end{aligned} \tag{3.11}$$

A network without all possible forward paths is known as sparsely connected network, or a non-fully connected network. The percentage of available connections that are utilized is known as the connectivity of the network.

3.6.2 Recurrent Neural Networks

A recurrent network can have connections that go backward from output nodes to input nodes and, in fact, can have arbitrary connections between any nodes. In this way, a recurrent network's internal state can alter as sets of input data are presented to it, and it can be said to have a memory. This is particularly useful in solving problems where the solution depends not just on the current inputs, but on all previous inputs. When learning, the recurrent network feeds its inputs through the network, including feeding data back from outputs to inputs through the network, and repeats this process until the values of the outputs do not change. At this point the network is said to be in a state of equilibrium or stability. A typical recurrent neural network can be explained by Fig. 3.9.

Hence, a recurrent network can be used as an error-connecting network. If only a few possible inputs are considered "valid", the network can correct all other inputs to the closest valid input.

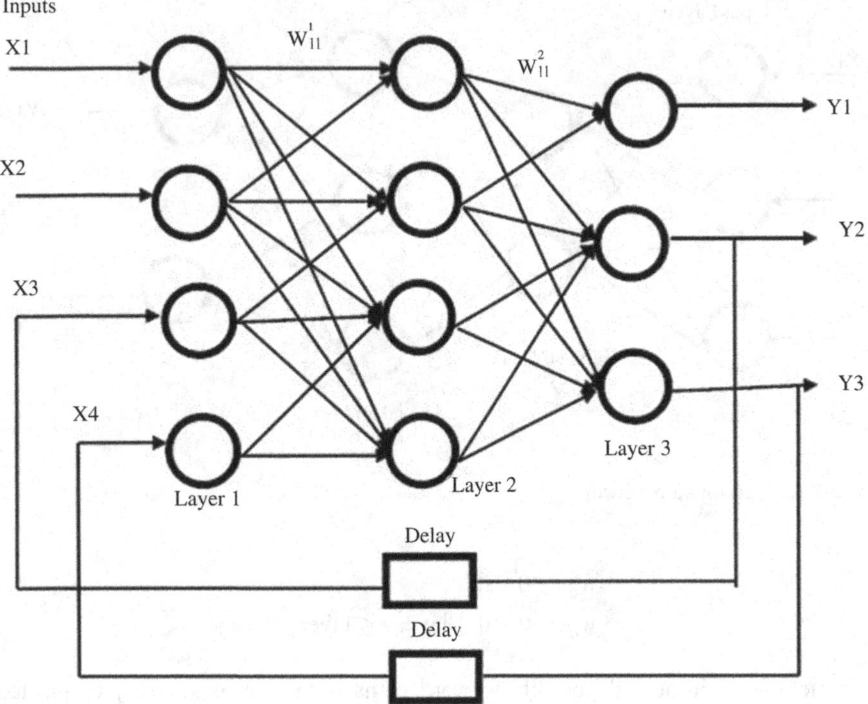

Fig. 3.9 A recurrent neural network

3.6.3 Radial Basis Function Neural Network

Radial basis function (RBF) network consists of three layers, input layer is first layer and basis function is the second layer as hidden layer and an output layer as shown in Fig. 3.10. Each node in the hidden layer represents a Gaussian basis function for all nodes and output node uses a linear activation function. Let W_{RBF}^k be the vector connection weight between the input nodes and the k-th RBF node or we can say $W_{RBF}^k = X - W^k$, so the output of the k-th RBF node is

$$h_{RBF}^k = \exp\left(-\frac{1}{\sigma_k^2}\left\|W_{RBF}^k\right\|^2\right) \tag{3.12}$$

where σ_k is the spread of k-th RBF function, $X = (x_1, x_2, \ldots, x_m)$ is the input vector, $W = (w_{1k}, w_{2k}, \ldots, w_{mk})$ and $\left|W_{RBF}^k\right|| = \sum_{i=1}^m (x_i - w_{ik})^2$. The output of the j-th output nodes can be computed as:

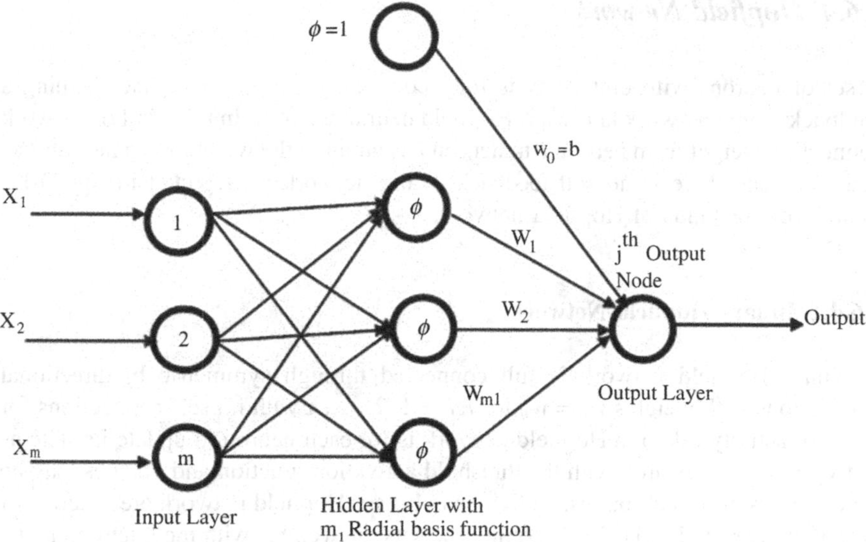

Fig. 3.10 Radial basis function network

$$O^j_{RBF} = \sum_{k=0}^{m_1} w_0^{kj} h^k_{RBF} \tag{3.13}$$

Training algorithm for RBF starts with one RBF node using one of the data points as the centre of the Gaussian functions, then it finds the data points with the highest error, which is used as the centre of a new RBF node. Squared errors are minimized by adjusting the connection weights between the hidden layer and the output layer. The process is continued till the error goal in terms of square of error is achieved as the number of RBF nodes attains a given maximum value. An RBF depends only on the distance to a centre point x_j and is of the form $\phi\|(x - x_j)\|$ and have a shape parameter ε in which $\phi(r)$ is replaced by $\phi(r, \varepsilon)$. Some of the most popular RBF's are:

(i) Piecewise smooth RBF's: $\phi(r)$
(ii) Piecewise polynomial (R_n): $|r^n|, n$ odd
(iii) Thin plate splines (TPSn): $|r^n|\ln|r|, n$ even
(iv) Infinitely smooth RBF's: $\phi(r, \varepsilon)$
(v) Multiquadric (MQ): $\sqrt{1 + (\varepsilon\, r)^2}$
(vi) Inverse multiquadric (IMQ): $\frac{1}{\sqrt{1+(\varepsilon\, r)^2}}$
(vii) Inverse quadratic (IQ): $\frac{1}{1+(\varepsilon\, r)^2}$
(viii) Gaussian (GS): $e^{-(\varepsilon\, r)^2}$

3.6.4 Hopfield Network

A set of neurons with unit delay is fully connected to each other and forming a feedback neural network known as Hopfield neural network. In this kind of network connection weight from neuron i to neuron j is equal to the weight from neuron j to neuron i and there is no self-feedback in the network as depicted in Fig. 3.11. There are four kinds of Hopfield network:

3.6.4.1 Binary Hopfield Network

A binary Hopfield network is full connected through symmetric bi directional connections with weights $w_{ij} = w_{ji}$ for $i, j = 1, 2, \ldots, n$ with no self connections for all i. An activity rule of a Hopfield network is for each neuron to update its state as if it were a single neuron with the threshold activation function and updates may be synchronous or asynchronous. Activities in binary Hopfield network are in terms of binary numbers (+1, −1). The learning rule sets the weights with the intention that a set of desired memories $\{x^{(t)}\}$ will be stable states of the Hopfield network's activity rule. The weights are set using the sum of outer products $w_{ij} = \eta \sum_n x_i^{(t)} x_j^{(t)}$, where η is an unimportant constant.

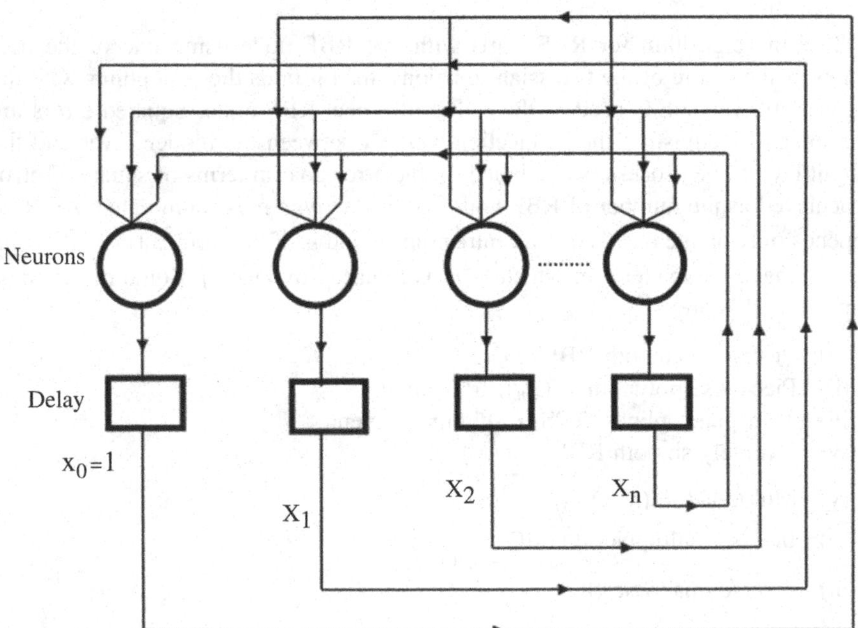

Fig. 3.11 Hopfield network

3.6.4.2 Continuous Hopfield Network

It uses the identical structure and learning rule of the binary Hopfield network but its activities are real numbers between −1 and +1. In continuous Hopfield network updates may be synchronous or asynchronous and involve equations

$$x_i = \sum_j w_{ij} x_j \quad \text{and} \quad a_i = \tanh(x_i)$$

or

$$a_i = \tanh(\beta x_i) = \frac{1 - e^{-\beta x_i}}{1 + e^{-\beta x_i}} \tag{3.14}$$

where $\beta \in (0, \infty)$ and β is known as the gain factor.

3.6.4.3 Continuous-Time Continuous Hopfield Network

In continuous time continuous Hopfield network x_i is continuous function of time t i.e. $x_i(t)$ which is computed as:

$$a_i(t) = \sum_j w_{ij} x_j(t) \tag{3.15}$$

And the response of a neuron to its activation is assumed to be mediated by the differential equation:

$$\frac{d}{dt}[x_i(t)] = \frac{1}{\tau}[x_i(t) - f(a_i)] \tag{3.16}$$

where, $f(a)$ is the activation function, such as $f(a) = \tanh(a)$. Every component of

$$\frac{d}{dt}(x_i(t))$$

has the same sign in case of continuous-time continuous Hopfield network, which means the system performs the steepest descent.

3.6.4.4 Discrete Hopfield Network

In the case of discrete Hopfield network the state vector $x(t)$ converges to a local error minimum for an initial state vector $x(0)$. The networks weight is given by the Hebb rule:

$$w_{ij} = \begin{cases} \frac{1}{n}\sum_{l=1}^{d} x_{li}x_{lj}, & i \neq j \\ 0, & i = j \end{cases} \tag{3.17}$$

The weight matrix is always symmetric and has zero diagonal elements. Activation functions are updates according to the given rule:

$$y_i(t) = f_{Hopfield}\left(x_i + \sum_{j \neq i}^{N} y_j(t-1)w_{ji}\right) \tag{3.18}$$

where t describes the dynamics of the activations of a collection of N neurons.

3.6.5 Cellular Neural Network

A cellular neural network is an artificial neural network which features a multi-dimensional array of neurons and local interconnections among cells. The important features of CNN paradigm are that it uses the analog cells with continuous signal values and local interaction within finite radius. Its architecture consists of regular spaced cloned circuits called cells, which is connected to its neighbor cells and can interact directly with each other. CNN consists of linear and non linear circuit elements, which typically are linear capacitors, linear resistors, linear and non linear controlled sources and independent sources as shown in Fig. 3.12, the typical circuit of a single cell. In the figure, $\mathrm{E}u_{ij}$ is the independent voltage source, I is the independent current source, I_n^{yij}, I_n^{uij} is voltage controlled current sources and E_{yij} is the output voltage controlled source. The cell has direct connections to its neighbors

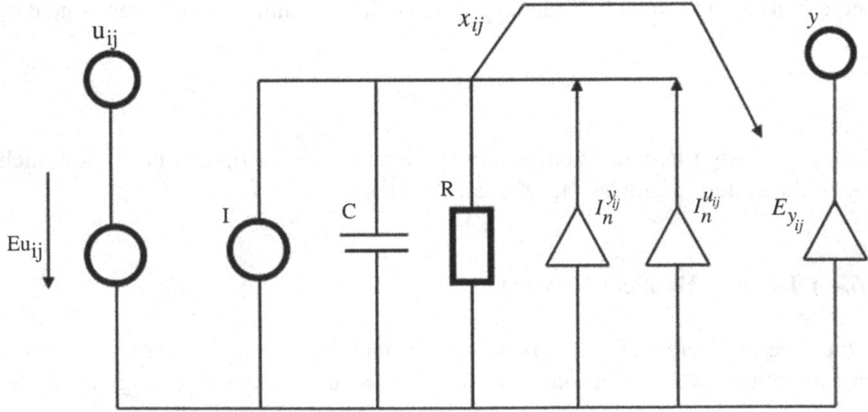

Fig. 3.12 Cellular neural network

through two kinds of weights: the feedback weight and the control weight and the index pair represent the direction of signal from one cell to another cell. The global behavior of CNN is characterized by a template set containing A-Template, B-Template, and the bias I. Cellular neural network has important potential applications in areas such as image processing and pattern recognition. It has the best features of both the words, its continuous time feature allows real time signal processing important in the digital domain and its local interconnection feature makes it tailor made for VLSI implementation.

3.6.6 Finite Element Neural Network

Finite element neural network represents the finite element model converted into the parallel network form. In the case with M elements and N nodes in the finite element mesh, the M network inputs take the α values in each element as input; N groups of N neurons are arranged in the hidden layer with N^2 neurons. The output of each group of hidden layer neurons is the corresponding row of the global matrix which is embedded by the material properties in each of elements. Each group of hidden neurons is connected to one output neuron by a set of weights φ, with each element of φ represents the nodal value ϕ_j as shown in Fig. 3.13 for a two element, four node FEM mesh. The output of each neuron is equal to b_i and each output neuron is a summation of linear activation function followed by a linear activation function:

$$b_i = \sum_{j=1}^{N} \phi_j \left(\sum_{k=1}^{M} \beta^k w_{ij}^k \right) \tag{3.19}$$

where β^k is the input corresponding to each element, w_{ij}^k is the weights from the input to the hidden layer. Figure 3.13 represents the FENN architecture with two input neurons, 16 hidden layer neurons and four output neurons. It represents the grouping of the hidden layer neurons and the similarity inherent in the weights that connect each group of hidden layer neurons to the corresponding output neuron.

3.6.7 Wavelet Neural Network

Wavelet neural network is an alternative to the feed forward neural network for approximating arbitrary non linear functions as an alternative. The basic idea for WNN is to replace the neurons by wavelons i.e. computing units obtained by cascading an affine transform and a multidimensional wavelet. Then these transforms and weights are identified from noise corrupted input/output data.

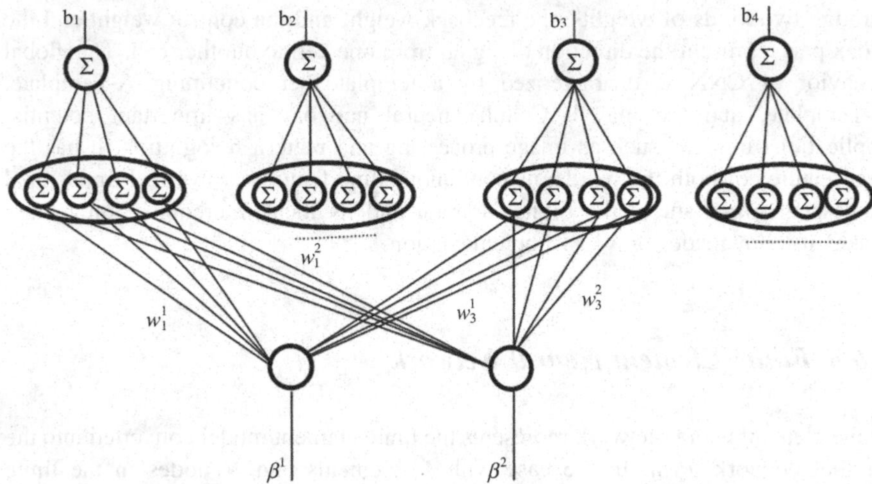

Fig. 3.13 Finite element neural network

It has the network structure of the form:

$$g(x) = \sum_{i=1}^{N} w_i \psi[D_i R_i (x - t_i)] + \bar{g} \tag{3.20}$$

where, the parameter $\bar{g}$ is used to approximate the function easily with non zero average, since the wavelet $\psi(x)$ is non zero mean average. The dilation matrices D_i's are diagonal matrices and R_i's are rotation matrices. Architecture of WNN is depicted in Fig. 3.14.

The initialization of the wavelet network consists in the evaluation of the parameters $\bar{g}, w_i, t_i$ and s_i for $i = 1, 2, \ldots, N$. $\bar{g}$ is initialized by estimating the mean

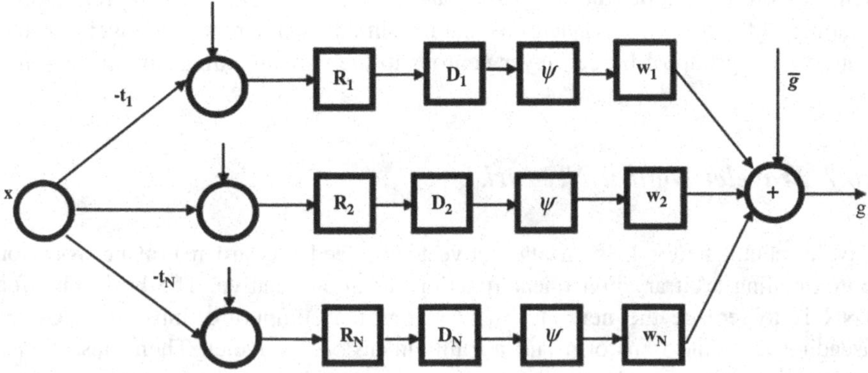

Fig. 3.14 Wavelet neural network

of function and set $\bar{g}$ to the estimated mean. Weights w_i's are set to zero and t_i's, S_i's are initialized by taking the point p between the domain the n set $t_1 = p$, $s_1 = \xi(b - a)$, where $\xi > 0$ is a properly selected constant. Interval is divided into the subintervals and t_2, s_2, t_3, s_3, and so on are initialized until all the wavelets are initialized. Point p is taken to be the centre of gravity of the domain $[a, b]$.

3.7 Learning in Neural Networks

A neural network has to be configured such that the application of a set of inputs produces the desired set of outputs. Various methods to set the strengths of the connection exist. One way is to set the weights explicitly, using priory knowledge. Another way is to train the neural network by feeding it, teaching patterns and letting it change its weights according to some learning rule. The term learning is widely used in the neural network field to describe this process; it might be formally described as: determining an optimized set of weights based on the statistics of the examples. The learning classification situations in neural networks may be classified into distinct sorts of learning: supervised learning, unsupervised learning, reinforcement learning and competitive learning.

3.7.1 Supervised Learning

A supervised learning method is one in which weight adjustments are made based on comparison with some target output. A teaching signal feeds into the neural network for the weight adjustments. These teaching signals are also known as training sample. A supervised learning algorithm is shown in following Fig. 3.15.

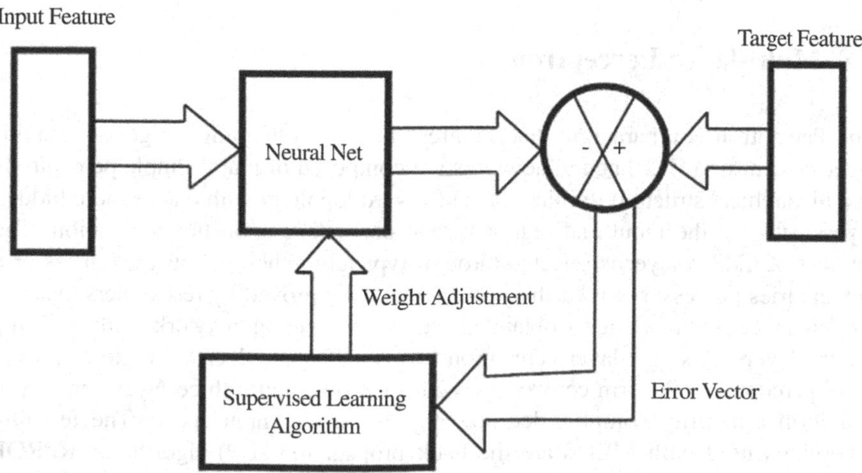

Fig. 3.15 A supervised learning algorithm

3.7.2 Unsupervised Learning

An Unsupervised learning method is one in which weight adjustments are not based on comparison with some target output. Here no teaching signal feeds into the weights adjustments but it requires some guidelines for successful learning. This property is also known as Self organization.

3.7.3 Reinforcement Learning

This learning requires one or more neurons at the output layer and the teacher or training sample, unlike supervised learning, indicates how closed the actual output is to the desired output. During the training session the error signal generated by the teacher is only binary, e.g. pass or fail, true or false, 1 or 0, in or out, etc. If the teacher's indication is fail, the network readjusts its parameters with the help of input signals and tries again and again until it gets its output response right i.e. pass.

3.7.4 Competitive Learning

A Competitive learning method is one in which several neurons are at the output layer of the network. When an input signal is applied to the network, each output neuron competes with the others to compute the closest output signal to the target. The network output for the applied input signals becomes the dominant one, and the remaining computed outputs from other neurons cease producing an output signal for that input signal.

3.8 Multi-layer Perceptron

The Perceptron is a paradigm that requires supervised learning. In general, multi-layer perceptron (MLP) neural networks are composed of many simple perceptrons in a hierarchical structure forming a feed forward topology with one or more hidden layers between the input and output layers, depending upon the application. The number of hidden layer is selected through typically either an educated guess or a cut and tries process, but it has been mathematically proved by researchers that one hidden layer is sufficient to obtain an equivalent neural network with multiple hidden layers. A single-layer perceptron forms half-plane decision regions, a two-layer perceptron can form convex (polygon) regions, and a three layer perceptron can form arbitrarily complex decision regions in the input space. The learning algorithms used with MLP's are the back propagation (BP) algorithms, RPROP learning algorithm, Levenberg-Marquardt learning algorithm, Genetic Algorithm and Particle swarm optimization.

3.8.1 Backpropagation Algorithm

A backpropagation method of ANN was first proposed by Werbos [27] in 1974. Later on Rumelhart, Hinton and Williams exploited backpropagation in their work in simulating cognitive process. Since then backpropagation has been employed in a number of fields for solving problems that would be quite difficult using conventional computer science techniques.

The backpropagation model of ANN has three layers of neurons: an input layer, a hidden layer, and an output layer, where there is no connection within the layer but fully connected between two consecutive layers. There are two synaptic (i.e. connection) weight matrices-one is in between the input layer and the hidden layer, and the other is in between the hidden layer and the output layer. There is a learning rate α in the subsequent formulae, indicating how much of the weight change should influence the current weight change. There is also a term indicating within what tolerance we can accept an output as 'good'. The backpropagation algorithm is an involved mathematical tool which has been widely used as a learning algorithm in feedforward multi-layer neural networks.

The main difficulty with MLP arises in calculating the weights of the hidden layers in an efficient way that results in the least output error. There is no direct observation of the error at the hidden layers. The weights are calculated during the learning phase of the network in this algorithm.

To apply the backpropagation learning procedure following is required:

(i) The set of normalized training patterns i.e., sample or data, both inputs $\{x_k\}$ and the corresponding targets $\{T_k\}$.
(ii) Value for the learning rate.
(iii) Criterion that terminates the algorithm.
(iv) Methodology for updating the weights, i.e., weight updating rules and different criteria for rule updation.
(v) Usually sigmoid activation is preferred by the user for non-linear activation.
(vi) Initial weight values (generally random numbers between −0.5 and 0.5).

3.8.2 The RPROP Learning Algorithm

RPROP stands for 'resilient propagation' and is an effective learning scheme that performs a direct adaption of weight parameters based on local gradient information. The RPROP algorithm [28, 29] was originally chosen as the gradient descent technique used as a learning algorithm due to its simplicity and the adaptive nature of its parameters. To explain this algorithm, consider a vector $u = [u_1, u_2, \ldots, u_n]$ which contains n number of weights to be optimized. Each weight u_i has an updated value Δ_i associated with it, which is added to and subtracted from the current weight value depending on the sign of derivative of $\partial E/\partial u_i$. Where $\partial E/\partial u_i$ represents the derivative of the error function with respect to network parameters and the error function which has to reduce can be written in the following form:

$$E = \frac{1}{|A|} \sum_{x \in A} F(x, N)^2 \tag{3.21}$$

Here, F is some signed error measure, A is a set of training points at which error is to be evaluated and N represents the output of the neural network. The weight of the parameters in the $(t+1)$th epoch is updated using the following scheme:

$$u_i^{t+1} = \begin{cases} u_i^t - \Delta_i^t, & if\left(\partial E/\partial u_i^{t+1}\right) > 0 \\ u_i^t + \Delta_i^t, & if\left(\partial E/\partial u_i^{t+1}\right) < 0 \end{cases} \tag{3.22}$$

The approach based on the above scheme is computationally inexpensive and very powerful since the updated values are adjusted dynamically rather than depending upon the magnitude of the derivative of the error function with respect to network parameters. The updated value in the $(t + 1)$-th epoch is adjusted according to the rule that if the derivative of the error function with respect to network parameters has same sign in the consecutive epochs indicates that the adjustment should be accelerated, and thus the current updated value is increased by a factor of $\eta^+ > 1$.

$$\Delta_i^{t+1} = \begin{cases} \eta^+ \Delta_i^t, & if \dfrac{\partial E}{\partial u_i^t} \cdot \dfrac{\partial E}{\partial u_i^{t+1}} > 0 \\ \eta^- \Delta_i^t, & if \dfrac{\partial E}{\partial u_i^t} \cdot \dfrac{\partial E}{\partial u_i^{t+1}} < 0 \end{cases} \tag{3.23}$$

Recommended value of $\eta^+ = 1.2$ is used, also if the sign of $\partial E/\partial u_i$ is changed in the next epoch then the minimum has been passed in the previous epoch and the updated value is reduced by a factor $0 < \eta^- < 1$. The update values and weights are changed every time the whole pattern set has been presented once to the network.

The main advantage of the RPROP algorithm is based on the fact that there is no need of the choice of parameters at all for many problems to obtain optimal convergences times. Also the algorithm is efficient with respect to both time and storage consumption and the other strength of this method is that convergence speed is not especially sensitive to the three parameter values $\eta^+ = 1.2$, $\eta^- = 0.5$ and $\Delta_0 = 0.1$. The only drawback of the method being the memory is required to store the $\partial E/\partial u_i^t$ and Δ_i^t values for each weight.

3.8.3 The Levenberg-Marquardt Learning Algorithm

Although RPROP algorithm is simple and computationally inexpensive still it fails to reduce error when solving more complicated boundary value problems. The Levenberg-Marquardt algorithm (LM) is also known as the damped least squares method is a classical approach which provides a numerical solution to the problem of minimizing a function over a space of parameters of the function [30]. It is known for its fast convergence using a sum of error squares error function as shown in Eq. (3.24). The Levenberg-Marquardt algorithm is derived by considering the error E after a differential change in the neural network weights from u_0 to u according to the second order Taylor series expansion

$$E(u) = E(u_0) + f^T(u - u_0) + \frac{1}{2}(u - u_0)^T H(u - u_0) + \cdots \tag{3.24}$$

where,

$$f = \nabla E(u) = \frac{\partial E}{\partial u} = \left[\frac{\partial E}{\partial u_1}, \frac{\partial E}{\partial u_2}, \ldots, \frac{\partial E}{\partial u_n}\right]^T \tag{3.25}$$

is the gradient vector and

$$H(u) = \begin{bmatrix} \frac{\partial^2 E(u)}{\partial u_1^2} & \frac{\partial^2 E(u)}{\partial u_1 \partial u_2} & \cdots & \frac{\partial^2 E(u)}{\partial u_1 \partial u_n} \\ \frac{\partial^2 E(u)}{\partial u_2 \partial u_1} & \frac{\partial^2 E(u)}{\partial u_2^2} & \cdots & \frac{\partial^2 E(u)}{\partial u_2 \partial u_n} \\ \cdot & \cdot & & \cdot \\ \cdot & \cdot & & \cdot \\ \cdot & \cdot & & \cdot \\ \frac{\partial^2 E(u)}{\partial u_n \partial u_1} & \frac{\partial^2 E(u)}{\partial u_n \partial u_2} & \cdots & \frac{\partial^2 E(u)}{\partial u_n^2} \end{bmatrix} \tag{3.26}$$

is the Hessian matrix. Taking the gradient of Eq. (3.24) with respect to the weight parameters the weight vector corresponding to the minimum error can be obtained:

$$\nabla E(u) = H(u^* - u_0) + f \tag{3.27}$$

Equating Eq. (3.27) to zero and solving for the weight vector with minimum error we can get:

$$u^* = u_0 - H^{-1} f \tag{3.28}$$

which results,

$$u_{i+1} = u_i - \eta_i H^{-1} f_i \tag{3.29}$$

where u_i is the weight vector in the i-th iteration and η is the learning rate and represents the Newton-Raphson learning algorithm. The error in Eq. (3.21) can be rewritten as:

$$E(u) = F(u)^T F(u) \tag{3.30}$$

where $F(u) = \left[F(x_1, \theta), F(x_2, \theta), \ldots, F\left(x_{|A|}, \theta\right)\right]^T$ given that $x_1, x_2, \ldots, x_{|A|} \in A$. The error is computed by summing the square of F at each location in the domain as determined by set of points. The Jacobian matrix is used to define the Hessian for the special case of sum of squared error.

$$H = 2J^T J + 2\frac{\partial J^T}{\partial u} F \tag{3.31}$$

where,

$$J(x, u) = \begin{bmatrix} \frac{\partial F(x_1,u)}{\partial u_1} & \frac{\partial F(x_1,u)}{\partial u_2} & \cdots & \frac{\partial F(x_1,u)}{\partial u_n} \\ \frac{\partial F(x_2,u)}{\partial u_1} & \frac{\partial F(x_2,u)}{\partial u_2} & \cdots & \frac{\partial F(x_2,u)}{\partial u_n} \\ \cdot & \cdot & & \cdot \\ \cdot & \cdot & & \cdot \\ \cdot & \cdot & & \cdot \\ \frac{\partial F\left(x_{\|A\|},u\right)}{\partial u_1} & \frac{\partial F\left(x_{\|A\|},u\right)}{\partial u_2} & \cdots & \frac{\partial F\left(x_{\|A\|},u\right)}{\partial u_n} \end{bmatrix} \tag{3.32}$$

The errors can be linearly approximated to produce $H \approx 2J^T J$, this approximation combined with Eq. (3.29) produces the Gauss Newton learning algorithm which assumes that the second hand term on Eq. (3.31) is negligible. The Levenberg–Marquardt method modifies the Gauss Newton algorithm by:

$$u_{i+1} = u_i - \frac{1}{2}\eta_i\left(J^T J_i + \lambda_i I\right)^{-1} f_i \tag{3.33}$$

where λ is scalar and I is the identity matrix. Large values are taken for learning rate η and factor λ in the beginning of the training and decreases as the solution improves. The Levenberg-Marquardt method has a fast convergence and effective optimization scheme for the weight parameters. The method is powerful and simple to operate after having few matrix operations.

3.8.4 Genetic Algorithm

Genetic algorithm searches the solution space of a function through the use of a simulated evolution i.e. the survival of the fittest strategy [31, 32]. It works on a set of elements of the solution space of the function that have to minimize. The set of elements is called a population and the elements of the set are called individual. The Genetic algorithm is applied to the problem which is a preventative of the optimization problem. The initial population could be defined randomly and or based on the prior knowledge. The algorithm will evaluate the individual of population based on the objective function and how much each agent is closed to the objective. To produce the next generation of people from the current generation, agents with better fitness are selected as the parents for the next generation and there are several operators which are applied to chromosomes to produce the next generations, these operators are known as genetic operators. Some important operators are Mutation, Crossover and Combination. So the new population is generated and its fitness will calculate and the process repeats up to maximum epoch assigned is achieved.

The procedure of Genetic algorithm can be written as the difference equation:

$$x[t+1] = s(v(x(t))) \tag{3.34}$$

where $x(t)$ is the population at time t, s is a selection operator and v is a random operator.

The shortcomings of the BP algorithm could be overcome if the training process is based on the global search of connection weights towards an optimal set defined by GA. For neural network to be fully optimal the learning rules are adapted dynamically according to its architecture and the problem. The basic learning rule can be given by the function:

$$\Delta w(t) = \sum_{k=1}^{n} \sum_{i_1, i_2, \ldots, i_k}^{n} \left(\theta_{i_1, i_2, \ldots, i_k = 1} \prod_{j=1}^{k} x_{ij}(t-1) \right) \tag{3.35}$$

where t is time, Δw is the weight change, θ's are the real value coefficients which will be determined by global search and $x_1, x_2, \ldots, x_n$ are local variables. Equation (3.35) is based on the assumption that same learning rules are applicable to every node of the network and weight updating is only dependent on the number of connection weights on a particular node. θ's are encoded as the real valued coefficient and the global search for learning rules can be done by the following rules:

(i) The evolution of the learning rules has to be implemented such that the evolution of architecture chromosomes is evolved at faster rate.
(ii) Fitness of the each evolutionary artificial neural network is evaluated.
(iii) Children for each individual in the current generation are reproducing using suitable selection method and depending upon the fitness.
(iv) Next generation is obtained by applying genetic operators to each individual child generated in the above 3rd rule.
(v) If the network has achieved the required error rate or the specified number of generations otherwise the procedure is repeated.

3.8.5 Particle Swarm Optimization

Particle swarm optimization is a non gradient based, probabilistic search algorithm which is based on a simplified social model and is inspired by the social behavior of animals and swarm theory [33, 34]. For optimizing the weight parameters in neural network the mean sum of square is defined in term of fitness function as:

$$F_j = \frac{1}{k_1}\sum_{i=1}^{k_1} (f^*())^2 + \frac{1}{k_2}\sum_{i=1}^{k_2} (Lf^*())^2 \quad for \quad j = 1, 2, 3, \ldots. \tag{3.36}$$

where k_1 is the number of steps, k_2 is the number of initial/boundary conditions, f^* is the algebraic sum of differential equation neural network representation that constitute a differential equation, L is the operator to define the initial/boundary conditions and j is the flight number. This fitness function representing in Eq. (3.36) has to be minimized using PSO for finding the global minimum from a huge space of the input data set.

The basic principle of particle swarm optimization is the adaptation between the individuals and this adaptation is a stochastic process that depends upon the memory of the each individual as well as the knowledge gained by the population as whole. The formulation of the problem in PSO is done by taking randomly generated particles called swarm as a population. This population of particles contains the randomly generated particle positions and velocity vectors for each particle. All particles in the swarm have fitness values and evaluated by fitness functions that is depends on the problem. The fitness of the function is defined by $f : R^n \to R$. The best way to update the initial position particles in PSO is to move towards its own direction, towards the globally best particle or towards the personally best particle. In each iteration, the position and velocity of the particle is updated according to its known previous local best position k_i^{n-1} and global best position of all particles k_g^{n-1} in the swarm so far. In PSO the updating formula for each particle velocity and position is given by:

$$v_i^n = uv_i^{n-1} + b_1\, rand()\left(k_i^{n-1} - x_i^{n-1}\right) + b_2\, rand()\left(k_g^{n-1} - x_i^{n-1}\right) \tag{3.37}$$

and

$$x_i^n = x_i^{n-1} + v_i^n \tag{3.38}$$

where u is the inertia weight which is linearly decreasing, rand() is the random number generated between 0 and 1, $i = 1, 2, 3, \ldots, m$, m is the number of particles in a swarm, b_1 and b_2 are the self confidence constants, n is the flight number. x_i^n and v_i^n are the position vector and velocity vector of the i-th particle of swarm at flight n respectively. In Particle swarm optimization, first of all initial population is randomly generated $K = [X_1, X_2, X_3, \ldots, X_n]$ and assign the velocities to each particle in the population $V = [V_1, V_2, V_3, \ldots, V_n]$, where K and V defines the initial population of sub swarms and n is the number of subpopulation. An initial population is generated in a bounded range with the random number generator in the following way:

$$\begin{aligned} x_j^i &= (B - A)^* r + A \\ v_j^i &= ((B - A)^* r + A)/2 \end{aligned} \tag{3.39}$$

for $j = 1, 2, 3, \ldots, m$, x_j^i is the j-th particle of the i-th sub swarm and v_j^i is the velocity of j-th particle of the i-th sub swarm. A and B represent the upper and lower bounds for the search dimension and r is a random number between 0 and 1. Then fitness function is evaluated and particles are ranked subject to the fitness value obtained also assign local best and global best accordingly. The position and velocity parameters are updated using Eqs. (3.37) and (3.38) until all the flights are achieved and store the global best particle. The distance of each stored global best particle is calculated by the following formula:

$$d_i = \left[\sum_{j=1}^{m} \left(|x_j| \right) \right] / m \tag{3.40}$$

for $i = 1, 2, 3, \ldots, N$. From the global best population $N/4$ particles are selected on the basis of maximum distance until the global best particles becomes equal to the number of particles in the subpopulation.

3.9 Neural Networks as Universal Approximator

Artificial neural network can make a non linear mapping from the inputs to the outputs of the corresponding system of neurons which is suitable for analyzing the problem defined by initial/boundary value problems that have no analytical solutions or which cannot be easily computed. One of the applications of the multilayer feed forward neural network is the global approximation of real valued multivariable function in a closed analytic form. Namely such neural networks are universal approximator. It has been find out in the literature that multilayer feed forward neural networks with one hidden layer using arbitrary squashing functions are capable of approximating any Borel measurable function from one finite dimensional space to another with any desired degree of accuracy. A squashing function is a function $f : R \rightarrow [0, 1]$ if it is non decreasing, $\lim_{\lambda \to \infty} f(\lambda) = 1$ and $\lim_{\lambda \to -\infty} f(\lambda) = 0$. To prove that multilayer feed forward neural networks as a class of universal approximator K. Hornik et al. presents various definitions and results in [35, 36].

Definition 3.1 If $A^r : R^r \rightarrow R$ is the set of all functions of the form $A(x) = w.x + b$, where w and x are vectors in R^r, $b \in R$ is a scalar and "." Is the usual dot product for any $r \in N$. In the context of neural network in the above definition x represents the input to neural network, w corresponds to network weights from input to indeterminate layer and b corresponds to the bias.

Definition 3.2 Let $\sum^r G$ be a class of functions for any Borel measurable function $G(.)$ is a mapping from R to R and $r \in N$.

$$f(x) = \sum_{j=1}^{q} O_j G(A_j(x)) \tag{3.41}$$

where, $x \in R^r, O_j \in R, A_j \in A^r$ and $q = 1, 2, \ldots$.

In this case $\sum^r G$ represents the class of output functions with squashing at the hidden layer and no squashing at the output layer and the scalars O_j corresponds to the network weight from hidden to the output layers.

Theorem 3.1 *If G be any continuous non constant function from R to R, then* $\sum\prod^r(G)$ *is uniformly dense on compacta in* C^r*, where* C^r *is a compact function from* R^r *to R. In other words* $\sum\prod$ *feed forward neural networks are capable of arbitrary accurate approximation to any real valued continuous function over a compact set. Another feature of this result is that the activation function should be any continuous non constant function.*

Theorem 3.2 *For every continuous non constant function G, every r and every probability measure* μ *on* (R^r, B^r), $\sum\prod^r G$ *is* ρ_u*-dense in* M^r*. Hence,* Theorem 3.2 *corresponds that the standard feed forward neural networks with only a single hidden layer can approximate any continuous function uniformly on compact set and any measurable function arbitrary well in* ρ_u *metric. Thus* $\sum$ *networks are also universal approximator.*

Theorem 3.3 *For every function g in* M^r *there is a compact subset K of* R^r *and an* $f \in \sum^r(\psi)$ *such that for any* $\varepsilon > 0$, $\exists\ \mu(k) < 1 - \varepsilon$ *and* $\forall x \in k, \exists\ |f(x) - g(x)| < \varepsilon$. *That defines that a single hidden layer feed forward neural network can approximate any measurable function to any desired degree of accuracy on some compact set of input patters that to the same degree of accuracy has measure. Thus the results established by Hornik et al. and given in the theorems proved that the neural networks are universal approximator that can approximate any Borel measurable set defined on a hypercube.*

Chapter 4
Neural Network Methods for Solving Differential Equations

Abstract In this chapter we presented different neural network methods for the solution of differential equations mainly Multilayer perceptron neural network, Radial basis function neural network, Multiquadric radial basis function network, Cellular neural network, Finite element neural network and Wavelet neural network. Recent development in all the above given methods has been also presented in this chapter to get better knowledge about the subject.

Keywords Multilayer perceptron · Radial basis function · Multiquadric functions · Finite element · Wavelet method · Cellular network

4.1 Method of Multilayer Perceptron Neural Network

Different neural network methods for the solution of differential equations are described in this chapter. For more details, we refer [37, 57, 63, 81, 102, 111]

A method based on MLP neural network has been presented in [37] for the solution of both ordinary differential equations (ODE's) and partial differential equations (PDE's). Method is based on the function approximation capabilities of feedforward neural networks and results in the construction of a solution which is in a differentiable and closed analytic form. This form employs a feedforward neural network as the basic approximation element, whose parameters (weights and biases) are adjusted to minimize an appropriate error function. Optimization techniques are used for minimizing the error quantity and training of the network, which in turn require the computation of error gradient with respect to the inputs and network parameters.

To illustrate the method, let us consider the problem of general differential equation to be solved as:

$$F\left(\vec{x}, y(\vec{x}), \nabla y(\vec{x}), \nabla^2 y(\vec{x})\right) = 0, \quad \bar{x} \in D \tag{4.1}$$

N. Yadav et al., *An Introduction to Neural Network Methods for Differential Equations*, SpringerBriefs in Computational Intelligence,
DOI 10.1007/978-94-017-9816-7_4

defined on certain boundary conditions where $\vec{x} = (x_1, x_2, \ldots, x_n) \in R^n$, $D \subset R^n$ denotes the definition domain and $y(\vec{x})$ is the solution to be computed. Following steps are required for the computation of the above differential equation Eq. (4.1)

(a) Transformation

First discretize the domain D and its boundary S into a set of discrete points $\hat{D}$ and $\hat{S}$ respectively. The problem is then transformed into a system of equations

$$F(\vec{x}_i, y(x_i), \nabla y(x_i), \nabla^2 y(x_i)) = 0 \quad \forall \quad \vec{x}_i \in \hat{D} \tag{4.2}$$

subject to the constraints imposed by the boundary conditions. If $y_t(\vec{x}, \vec{p})$ denotes a trial solution with the adjustable parameters $\vec{p}$, the problem is transformed to an optimization problem

$$\min_{\vec{p}} \sum_{x_i \in \vec{D}} F((\vec{x}_i), y_t(\vec{x}_i, \vec{p}), \nabla y_t(\vec{x}_i, \vec{p}), \nabla^2 y_t(\vec{x}_i, \vec{p}))^2 \tag{4.3}$$

subject to the constraints imposed by the boundary conditions.

(b) Construction of Trial Solution

To construct the trial function $y_t(\vec{x})$ we assume that the trial function satisfies the given boundary conditions and it is the sum of two terms-one is independent of adjustable parameters, and the other is with adjustable parameters. Suppose the trial function is

$$y_t(\vec{x}) = A(\vec{x}) + f(\vec{x}, N(\vec{x}, \vec{p})) \tag{4.4}$$

where, $A(\vec{x})$ contains no adjustable parameters which satisfies the initial/boundary conditions and $N(\vec{x}, \vec{p})$ is a single output feed forward neural network with parameters $\vec{p}$ and n input feds with the input vector $\vec{x}$. The second term f is constructed in a way so that it does not contribute to the boundary conditions, since $y_t(\vec{x})$ must also satisfy them. This term represents a neural network whose parameters are to be adjusted in order to solve minimization problem, hence the problem has been reduced to the unconstrained optimization problem from the original constrained one which is much easier to handle due to the choice of the form of the trial solution that satisfies by construction the boundary conditions.

4.1.1 Gradient Computation

Minimization of error function can also be treated as a procedure of training the neural network, where the error corresponding to each input vector $\vec{x}_i$ is the value $f(\vec{x}_i)$ which has to become zero. In Computation of this error value, it requires the network output as well as the derivatives of the output with respect to the input vectors. Therefore, while computing error with respect to the network parameters, we need to compute not only the gradient of the network but also the gradient of the network derivatives with respect to its inputs.

4.1.2 Gradient Computation with Respect to Network Inputs

Next step is to compute the gradient with respect to input vectors, for this purpose let us consider a multilayer perceptron (MLP) neural network with n input units, a hidden layer with H sigmoid units and a linear output unit. For a given input vector $\vec{x} = (x_1, x_2, \ldots, x_n)$ the output of the network can be given as:

$$N = \sum_{i=1}^{H} v_i \sigma(z_i) \tag{4.5}$$

where,

$$z_i = \sum_{j=1}^{n} w_{ij} x_j + u_i,$$

In Eq. (4.5) w_{ij} denotes the weight from the input unit j to the hidden unit i, v_i represents weight from the hidden unit i to the output, u_i is the bias of hidden unit i, and $\sigma(z)$ is the sigmoid activation function. Now the derivative of networks output N with respect to input vector x_j is:

$$\frac{\partial N}{\partial x_j} = \frac{\partial}{\partial x_j}\left(\sum_{i=1}^{H} v_i \sigma\left(\sum_{j=1}^{n} w_{ij} x_j + u_i\right)\right) = \sum_{i=1}^{h} v_i w_{ij} \sigma^{(1)} \tag{4.6}$$

where,

$$\sigma^{(1)} = \frac{\partial \sigma(x)}{\partial x}$$

Similarly, the k-th derivative of N is

$$\frac{\partial^k N}{\partial x_j^k} = \sum v_i w_{ij}^k \sigma_i^{(k)} \tag{4.7}$$

where, $\sigma_i = \sigma(z_i)$ and $\sigma^{(k)}$ denotes the k-th order derivative of the sigmoid activation function. In general the derivative for any order with respect to any of input vector can be given as:

$$\frac{\partial^{\lambda_1}}{\partial x_1^{\lambda_1}}\frac{\partial^{\lambda_2}}{\partial x_2^{\lambda_2}}\cdots\frac{\partial^{\lambda_n}}{\partial x_2^{\lambda_n}}N=\sum_{i=1}^{n}v_iP_i\sigma_i^{(\Lambda)}=N_k(\vec{x}) \tag{4.8}$$

and

$$P_i=\prod_{k=1}^{n}w_{ik}^{\lambda_k}\quad \Lambda=\sum_{i=1}^{n}\lambda_i \tag{4.9}$$

4.1.3 Gradient Computation with Respect to Network Parameters

Network's derivative with respect to any of its inputs is equivalent to a feed-forward neural network $N_k(\vec{x})$ with one hidden layer, having the same values for the weights w_{ij} and thresholds u_i and with each weight v_i being replaced with v_ip_i. Moreover, the transfer function of each hidden unit is replaced with the Λ-th order derivative of the sigmoid function. Therefore, the gradient of N_k with respect to the parameters of the original network can be easily obtained as:

$$\frac{\partial N_k}{\partial v_i}=P_i\sigma_i^{(\Lambda)} \tag{4.10}$$

$$\frac{\partial N_k}{\partial u_i}=v_iP_i\sigma_i^{(\Lambda+1)} \tag{4.11}$$

$$\frac{\partial N_k}{\partial w_{ij}}=x_jv_iP_i\sigma_i^{\Lambda+1}+v_i\lambda_jw_{ij}^{\lambda_j-1}\left(\prod_{k=1,k\neq j}w_{ik}^{\lambda_k}\right)\sigma_i^{(\Lambda)} \tag{4.12}$$

4.1.4 Network Parameter Updation

After computation of derivative of the error with respect to the network parameter has been defined then the network parameters updation rule can be given as,

$$v_i(t+1)=v_i(t)+\alpha\frac{\partial N_k}{\partial v_i} \tag{4.13}$$

$$u_i(t+1)=u_i(t)+\beta\frac{\partial N_k}{\partial u_i} \tag{4.14}$$

$$w_{ij}(t+1) = w_{ij}(t) + \gamma \frac{\partial N_k}{\partial w_{ij}} \tag{4.15}$$

where α, β and γ are the learning rates, $i = 1, 2, \ldots, n$ and $j = 1, 2, \ldots, h$.

Once a derivative of the error with respect to the network parameters has been defined it is then straightforward to employ any optimization technique to minimize error function.

Remark 1 The study of the above method presented in [37] concludes that it can be applied to both ODE's and PDE's by constructing the appropriate form of the trial solution. Presented method also exhibits excellent generalization performance as the deviation at the test points was in no case greater than the maximum deviation at the training points. It can also be stated that the method can easily be used for dealing with domains of higher dimension.

4.1.5 Recent Development in MLPNN for Solving Differential Equations

Multilayer perceptron neural network method with the extended back propagation algorithm is presented by He et al. in [38] to train the derivative of a feed forward neural network. They presented a method to solve a class of first order partial differential equation as input to state linearizable or approximate linearizable systems and to examine the advantage of the method.

4.1.5.1 Extended Back Propagation Algorithm

For training a feed forward network to map a set of n dimensional input/m dimensional output vector pairs (x_i, T_i) for $i = 1, 2, \ldots, m$ are considered as the problem of nonlinear least squares. If the output of an n layered feed forward neural network is o_i^n for input vector x_i, then the error function for nonlinear least squares can be defined as:

$$E = \sum_{i=1}^{m} \left(T_i - o_i^n\right)^T \left(T_i - o_i^n\right)$$

A general structure for calculating output can be given as:

The extended back propagation algorithm for training the derivative of the network can be given as: If we have input x_i and output o_i, the performance index of the network can be given by:

$$E = \frac{1}{2}\sum_{i=1}^{m}\left(o_i^T - G_i\right)^T\left(o_i^T - G_i\right) \tag{4.16}$$

where, $G(x)$ represents the transpose of the output of the network with respect to input x. Gradient of the output of the network with respect to the output in k-th layer will be:

$$\left(\frac{\partial O^n}{\partial O^k}\right)^T = \Delta^k \tag{4.17}$$

then, $\Delta^{n-1} = w^{nT}F^n(b^n)$, $\Delta^{n-2} = w^{(n-1)T}F^{n-1}(b^{n-1})\Delta^{n-1}$ and similarly $\Delta^0 = w^{1T}$ $F^1(b^1)\Delta^1$. Thus $G(x)$ can be represented as $G(x) = \Delta^0 + b^{0T}$. Then the derivative of the squared errors of a single input/output pair $\hat{E} = (o_i^T - G_i)^T(o_i^T - G_i)$ with respect to Δ^k, weights and biases are computed. Simulation technique is used to demonstrate the effectiveness of the proposed algorithm and it is shown that the method can be very useful for practical applications in the following cases:

(i) When a non linear system satisfies the conditions for input-to-state linearization but the nontrivial solution of the given equation

$$\frac{\partial \lambda(x)}{\partial x}\left[g(x)\ ad_f^1 g(x) \ldots ad_f^{n-2} g(x)\right] = 0 \tag{4.18}$$

is hard to find by training the derivative of a neural network. We can seek the approximate solution by the method given in [38]. Since there is no restrictive condition for choosing the basis vector to train the neural network, therefore a simple transformation to construct the basis is recommended

(ii) When a nontrivial solution does not exist for the above given equation, we still obtain an approximate solution. If the approximate solution is considered an exact solution for a linearizable feedback system, then the system should approximate the given non linear system as closely as possible. The extended backpropagation algorithm can benefit the design of the class of nonlinear control systems, when the non trivial solution of partial differential equations is difficult to find. The control design based on this method cannot result in a satisfactory way for the applications where a large region of operation is required.

In Lagaris et al. [39] presented an artificial neural network with the synergy of the two feed forward networks of different types to solve the partial differential equation.

4.1.5.2 Model Based on MLP-RBF Synergy

The constrained optimization problem may be tackle in a way such that constraints are exactly satisfies by construction of a model and to use a suitable constrained minimization problem for the constraints. A model has been developed with the combination of feedforward and RBF networks as:

$$\Psi_M(\vec{x},p) = N(\vec{x},p) + \sum_{l=1}^{M} q_l e^{-\lambda|\vec{x}-\alpha\vec{r}_l+\vec{h}|^2} \tag{4.19}$$

where, the first term represents a multilayer perceptron with p representing the set of its weights and biases and second term represents an RBF network with M hidden units that all share a common exponential factor λ. $||$ denotes the Euclidean norm, the coefficients q_l are uniquely determined by requiring that the boundary conditions are satisfied i.e. if we consider a partial differential equation of the form:

$$L\psi = f \tag{4.20}$$

Together with the boundary conditions defined on the M points inside the boundary as:

$$\psi(\vec{r}_i) = b_i \quad \text{(Dirichlet)} \tag{4.21}$$

or

$$\hat{n}_i \cdot \nabla\, \psi(\vec{r}_i) = c_i \quad \text{(Neumann)} \tag{4.22}$$

where L is a differential operator and $\psi = \psi(\vec{x})$ $(\vec{x} \in D \subset R^{(n)})$ with Dirichlet or Neumann boundary conditions. The boundary $B = \partial D$ can be any arbitrary complex geometrical in shape. $\hat{n}_i$ is the outward unit vector, normal to boundary at the point $\vec{r}_i$. Collocation method is used to prepare the energy function for the minimization process as:

$$\min_p\, E(p) = \sum_{i=1}^{K} (L\psi_M(\vec{x}_i,p) - f(\vec{x}_i))^2 \tag{4.23}$$

Subject to the constraints imposed by the boundary conditions. The coefficients q_l are determined by the equation:

$$b_i - N(\vec{r}_i,p) = \sum_{i=1}^{M} q_l\, e^{-\lambda|\vec{r}_i-\alpha\vec{r}_i+\vec{h}|^2} \tag{4.24}$$

or,

$$c_i - \vec{n}_i \cdot \nabla N(\vec{r}_i, p) = -2\lambda \sum_{i=1}^{M} q_l e^{-\lambda|\vec{r}_i - \alpha\vec{r}_i + \vec{h}|^2} \vec{n}_i \cdot (\vec{r}_i - \alpha\vec{r}_i + \vec{h}) \tag{4.25}$$

for Dirichlet and Neumann boundary conditions respectively. Therefore a set of linear systems have to solve for obtaining the coefficient in both the cases or penalty function method can be used to minimize the problem. The model based on the combination of MLP and RBF satisfies the boundary condition exactly but it is computationally expensive since one has to solve a system of linear equations at every evaluation of the model. Penalty method is efficient but it does not satisfy the boundary conditions exactly, hence the combination of both these methods is used by the authors. Penalty method is used to obtain a model that satisfies the boundary condition approximately and then refines using the synergy method. Solutions obtained by the given approach shows that the method is equally effective, and retains its advantage over the Galerkin Finite element method. It also provides accurate solutions in a closed analytic form that satisfy the boundary conditions at the selected points.

In [40], the authors described a method that involves the combination of artificial neural networks and evolutionary algorithm to solve partial differential equation and its boundary or initial conditions. They used the concept that multiple input, single output and single hidden layer feed forward networks with a linear output layer with no bias are capable of approximating arbitrary functions and its derivatives.

4.1.5.3 MLP with Evolutionary Algorithm

To clarify the working of the method, following differential equation has been taken with two of its initial conditions:

$$\frac{d^2y}{dt^2} + y = 0, \quad t \in [0, 1] \tag{4.26}$$

$$\frac{dy}{dt} = 1, \quad y(0) = 0 \tag{4.27}$$

By assuming that

$$y, \frac{dy}{dt}, \frac{d^2y}{dt^2}$$

are continuous mappings, these are approximated by the log sigmoid function mappings arbitrary as:

$$\phi(t) = \sum_{i=1}^{m} \alpha_i f(w_i t + \beta_i) \tag{4.28}$$

$$\frac{d\phi}{dt} = \sum_{i=1}^{m} \alpha_i w_i \frac{df}{dt}(w_i t + \beta_i) \tag{4.29}$$

$$\frac{d^2\phi}{dt^2} = \sum_{i=1}^{m} \alpha_i w_i^2 \frac{d^2 f}{dt^2}(w_i t + \beta_i) \tag{4.30}$$

A differential equation neural network is then constructed having five layers with bias in the first two layers. All obtained networks are trained simultaneously as a consequence of their interrelationship using evolutionary algorithm for calculating the solution of the partial differential equation and its boundary and initial conditions. To apply the evolutionary algorithm the mean sum of squares errors for each training set of differential equation neural network is defined which represents the square of the difference between the target and the output of the network which is summed for all inputs and that sum is divided by the number of inputs. So the following expression:

$$e_1 + e_2 + e_3 \tag{4.31}$$

is minimized for the above case by using an evolutionary algorithm using the GEATbx toolbox in Matlab. It has been observed by the authors that good results can be obtained if they restrict the values of the variables to the interval [−5, 5]. The knowledge about the partial differential equations and its boundary and/or initial conditions has been incorporated into the structures and the training sets of several neural networks and found that the results for one and two dimensional problem are very good in respect of efficiency, accuracy, convergence and stability.

Smaoui and Al-Enezi in [41] presented combination of Karhunen-Loeve (K-L) decomposition and artificial neural networks to analyze the dynamics of two non linear partial differential equations known as the Kuramato-Sivashinsky (K-S) equation and the two dimensional Navier-Stokes (N-S) equation.

4.1.5.4 MLP with K-L Decomposition

The K-S equation is a model equation for interfacial instabilities in the terms of angular phase turbulence for a system of reaction diffusion equation that models the Belouzov-Zabotinskii reaction in three space dimensions and can be given as:

$$\frac{\partial y}{\partial t} + v\frac{\partial^4 y}{\partial x^4} + \frac{\partial^2 y}{\partial x^2} + \frac{1}{2}\left(\frac{\partial y}{\partial x}\right)^2 = 0 \tag{4.32}$$

Together with the conditions:

$$y(x,t) = y(x+L,t) \tag{4.33}$$

$$y(x,0) = y_0(x) \tag{4.34}$$

The time series solution of Eq. (4.32) is computed using $y_0(x) = \sin 2x$ by decomposing $y(x,t)$ as:

$$y(x,t) = \sum_{k=-\infty}^{\infty} a_k(t)e^{ikx} \tag{4.35}$$

The problem described in Eqs. (4.32–4.34) are solved using a pseudo spectral Galerkin method where the nonlinear term is treated using a "de-aliasing" technique known as aliasing removal by truncation. The numerical solution obtained using the technique consists of two laminar states: one between two bursts and the other is on the other sides of the two bursts for $\alpha = 17.75$. K-L decomposition was applied on the numerical simulation data to extract coherent structures of the dynamical behavior represented by heteroclinic connection. A neural network model is then constructed with one input layer, two hidden layers both with log sigmoid activation functions and an output layer. The input layer consists of five data coefficients' at time t_n and t_{n-1} and output is the following mapping:

$$a_i(t_n + P) = f(a_i(t_n), a_i(t_{n-1})), \quad i = 1, \ldots, 5 \tag{4.36}$$

where f is the set of non linear function that represent the neural network model. Network is trained for the data set and when the sum square error reaches the preset bound, the weights connecting all the bounds are saved and the network is again trained for testing a new set of data coefficients. For modeling and prediction of P times steps into the future dynamical behavior artificial neural network is used at $\alpha = 17.75$ for different values of P. Authors found that the neural network model is able to capture the dynamics of system, and observed that as P increases, the model behavior degrades. Eight different tori were obtained while applying the symmetry observed in the two-dimensional N-S equations on the quasiperiodic behavior. They showed that by exploiting the symmetries of the equation and using K-L decomposition in conjunction with neural networks, a smart neural model can be obtained.

Malek and Beidokhti in [42] presented a novel hybrid method for the solution of high order ordinary differential equations.

4.1.5.5 MLP with Nelder-Mead Simplex Method

The hybrid technique adopted here as a combination of the neural network and the Nelder-Mead optimization technique nicely produce non linear function named as energy function using neural network and minimization of the function is guaranteed with the optimum structure of the non linear solution function. Let us consider a general initial/boundary value problem of the form:

$$\begin{cases} F\left[x, y(x), \dfrac{dy}{dx}, \dfrac{d^2y}{dx^2}, \ldots, \dfrac{d^ny}{dx^n}\right] = 0, & x \in [a, b] \\ B\left[x, y(x), \dfrac{dy}{dx}, \dfrac{d^2y}{dx^2}, \ldots, \dfrac{d^ny}{dx^n}\right] = 0, & x = a \text{ and/or } b \end{cases} \tag{4.37}$$

where F is a differential operator of degree n, B is an initial/boundary operator, y is an unknown dependent variable to be calculated and x is an independent variable belonging to $[a, b]$. Solution to Eq. (4.37) is of the form $y_T(x, P)$ where y_T is a dependent variable to x and P, and P is an adjustable parameter involving weights and biases in the structure of three layer feed forward neural networks which satisfies the following optimization problem:

$$\begin{cases} \underset{F}{\text{Min}} \displaystyle\int_a^b \left\| D\left[x, y_T(x,P), \dfrac{dy_T(x,P)}{dx}, \ldots, \dfrac{d^ny_T(x,P)}{dx^n}\right] \right\|^2 dx \\ B\left[x, y_T, \dfrac{dy_T(x,P)}{dx}, \ldots, \dfrac{d^ny_T(x,P)}{dx^n}\right] = 0, \quad x = a \text{ and/or } b \end{cases} \tag{4.38}$$

In order to deal with Eq. (4.38) it is simpler to deal the following constrained optimization problem of the form

$$y_T(x, P) = \alpha(x) + \beta[x, N(x, P)] \tag{4.39}$$

where, first term involves adjustable parameters and satisfies initial or boundary conditions and second term represents three layered feed forward neural network. Minimization in Eq. (4.38) has been done which is considered as a training process for the proposed neural network and the error $E(x)$ corresponding to every entry x has to be minimized. A three layered perceptron with single entry, one hidden layer and one unit output is considered for training procedure. Authors used Nelder-Mead simplex method [43] to compute the error $E(x)$ from substitution of $y_T(x, w, v, b)$ into the Eq. (4.37). It has been concluded that the proposed method act as a good interpolation as well as an extrapolation method for calculating the close enough point outside the boundary points of the interval.

In article [51], the authors solved the first order initial value problem in ordinary differential equations using cosine function as the transfer function of neural network.

4.1.5.6 MLP Based on Cosine-Based Model

The model based on the cosine basis function as a transfer function of the hidden layer neuron has been presented and the model of neural network based on cosine basis function can be represented by Fig. 4.1 as:

$$c_n(x) = \sum_{n=0}^{N-1} \cos(nx)$$

Here w_n is weight vector.

This equation is a transfer function of the hidden layer neuron, $x \in [0, \pi]$, weight matrix vector is $W = [w_0, w_1, \ldots, w_{n-1}]^T$ and transfer function matrix as

$$c(x) = (c_0(x), c_1(x), \ldots, c_{n-1}(x))^T,$$

N is the number of hidden layer neurons and Neural network output is given as

$$\hat{y}(x) = \sum_{n=0}^{N-1} \omega_n \cos\, nx \tag{4.40}$$

A neural network algorithm is developed by considering the initial value problems in ordinary differential equations, for which the error function is:

$$e(k) = \sum_{n=1}^{N-1} n\, \omega_n \, \sin\, nx - \beta + f(x, \hat{y}(x)) \tag{4.41}$$

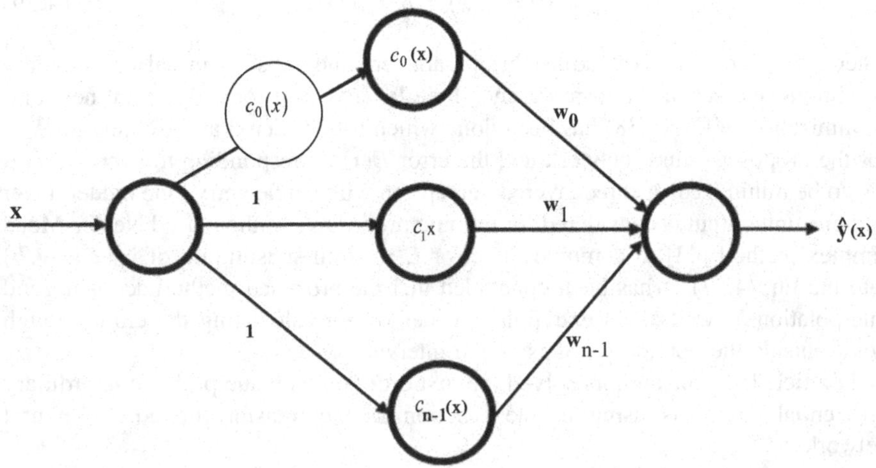

Fig. 4.1 Neural Network based on cosine based model

and weights are adjusted as;

$$
\begin{aligned}
w_n(k+1) = w_n(k) &= -\mu\, e(k)\left[n \sin\, nx_k + f_y(x, \hat{y}(x_k) \cos\, nx_k)\right] \\
n &= 0, 1, \ldots, N-1
\end{aligned}
\tag{4.42}
$$

where μ is learning rate and $0<\mu<1$.The convergence criterion for the cosine based neural network is then given by the theorem as: Suppose that μ is the learning rate, N is the number of hidden layer neurons, L is the upper limit of

$$\frac{\partial f(x,y)}{\partial y}$$

Then

$$\left|\frac{\partial f(x,y)}{\partial y}\right| \leq L,$$

If learning rate satisfies

$$0<\mu<\frac{12}{N(2N^2+6L^2-3N+1)}$$

then the algorithm is convergent.

The algorithm based on cosine basis function is more precise and provides a new approach on numerical computation that result at an arbitrary x between two adjacent nodes.

In [44] authors presented the numerical solution of nonlinear Schrodinger equation by feed forward neural network and the improvement of energy function is done by the unsupervised training method.

4.1.5.7 MLP with Unsupervised Training Network

Authors considered the following time independent Schrodinger equation of the motion of a quantum particle in a one dimensional system:

$$\hat{H}\psi(x) \equiv \left(-\frac{h^2}{2m}\frac{\partial^2}{\partial x^2} + V(x)\right)\psi(x) = E\psi(x) \tag{4.43}$$

where, m and $V(x)$ are the mass of a particle and the potential function, respectively. $\hat{H}$, $\psi(x)$, and E denote the system Hamiltonian, Eigen function and the Eigen value, respectively. The identity of quantum mechanical state can be represented by the wave function $\psi(x) = A(x) \cdot S(x)$ in the coordinate system. Hence the energy function:

$$E^q = |\frac{\partial^2 \psi(x)}{\partial x^2} + E\psi(x)|^2 + |\sum_{k=1}^{2} C_k|^2 \tag{4.44}$$

Since the eigen value parameter is unknown in the Eq. (4.44), its value is initialized and the network is then trained with specified hidden units. If the energy function E^q does not converge to zero, eigen values are updated and tries again. The algorithm for the above method with unsupervised training methodology is given by the authors as in Fig. 4.2. Main goal of the algorithm is that the energy function E^q must be zero, if it does not converge to zero, after finishing these cycles hidden layers are increased and then tries again. The authors represented the wave function by the feed forward artificial neural network, in which the coordinate value is regarded as an input while the networks output are assigned to two separate parts. They obtained energy function of the artificial neural network from the Schrodinger equation and its boundary conditions and used unsupervised training method for training the network.

Accuracy of the method is shown by comparing the results to the results that are analytically known and also by the Runge-Kutta method of order four. The method can be used for domains with higher dimension also.

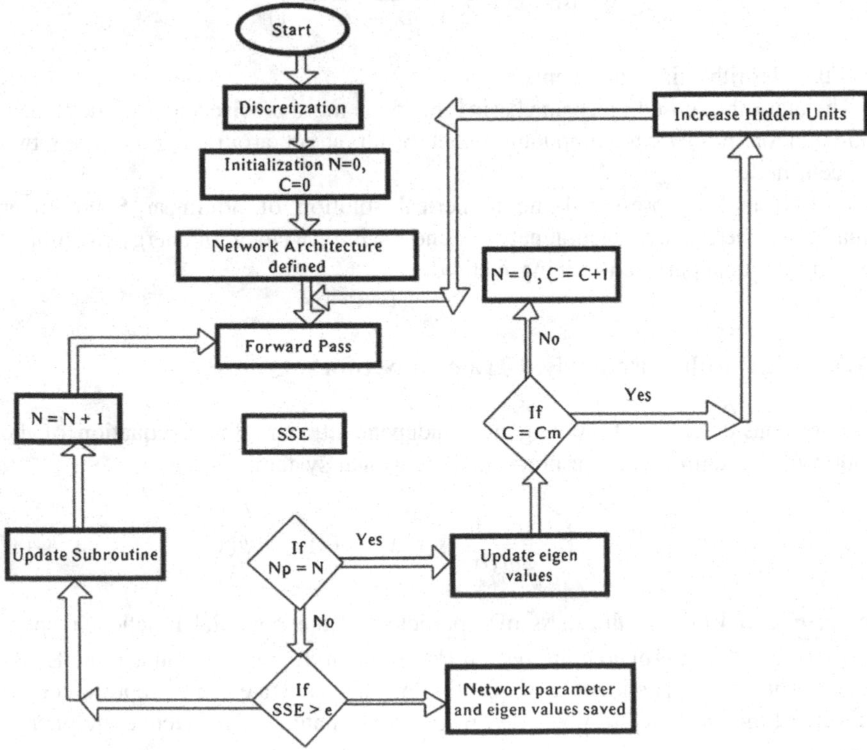

Fig. 4.2 An algorithm for neural network with unsupervised learning

In [45], the author proposed a hybrid method based on artificial neural networks, minimization techniques and collocation method to determine a related approximate solution in a closed analytic form of the time independent partial differential equations.

4.1.5.8 MLP for Time-Dependent Differential Equations

To describe the artificial neural network method for time dependent differential equations a set of initial/boundary value problem for time dependent equations has been taken of the form:

$$\begin{cases} \forall i_1 = 1,\ldots,p_1 : D_{i_1}\left[t,x,\ldots,\dfrac{\partial^{\alpha_0+\alpha_1+\cdots+\alpha_n}}{\partial t^{\alpha_0}\partial x_1^{\alpha_1}\cdots\partial x_n^{\alpha_n}}y_i(t,x),\ldots\right] = 0, & t\in[t_0,t_{\max}], \quad x\in\Omega \\ \forall i_2 = 1,\ldots,p_2 : I_{i_2}\left[t_0,x,\ldots,\dfrac{\partial^{\alpha_0+\alpha_1+\cdots+\alpha_n}}{\partial t^{\alpha_0}\partial x_1^{\alpha_1}\ldots\partial x_n^{\alpha_n}}y_i(t_0,x),\ldots\right] = 0, & x\in\Omega, \quad 1\leq i\leq m \\ \forall i_3 = 1,\ldots,p_3 : B_{i_3}\left[t,x,\ldots,\dfrac{\partial^{\alpha_0+\alpha_1+\cdots+\alpha_n}}{\partial t^{\alpha_0}\partial x_1^{\alpha_1}\ldots\partial x_n^{\alpha_n}}y_i(t,x),\ldots\right] = 0, & t\in[t_0,t_{\max}], \quad x\in\Omega \end{cases} \tag{4.45}$$

where the real valued multivariable functions D_{i_1}, I_{i_2} and B_{i_3} represents the known and non linear time dependent system of partial differential equations, initial and boundary conditions, respectively, t is the time variable, x is the real valued spatial variable, $\Omega \subseteq R^n$ is a bounded domain, $(\alpha_0,\alpha_1,\ldots,\alpha_n)\in N_0^{n+1}(N_0 = N\cup\{0\})$ is multi index variable. A trial approximate solution has been prepared as:

$$y_T(t,x,P) = [y_{T_1}(t,x,P_1),\ldots,y_{T_m}(t,x,P_m)] \tag{4.46}$$

which includes m three layered feedforward neural networks and contains adjustable parameters (weights and biases) for the solution. To obtain the proper values of the adjustable parameters the problem is transformed into unconstrained optimization problem. For any desired order differentiable functions $\gamma_i : R^{n+2}\to R(i = 1,\ldots,m)$ the trial solution can be assumed as:

$$y_{T_i}(t,x,P) = \gamma_i[t,x,N_i(t,x,P)] \tag{4.47}$$

In the case of mixed partial derivatives for the input values $(t,x_1,x_2,\ldots,x_n)$ the output of the network is:

$$N(t,x_1,x_2,\ldots,x_n) = \sum_{i=1}^{H} v_i\, s\left[w_i t + \left(\sum_{j=1}^{n} w_{ij}x_j\right) + b_i\right] \tag{4.48}$$

where, v_i is the synaptic weight from the i-th hidden neuron to the output, w_i is the synaptic coefficient from the time input to the i-th hidden neuron, w_{ij} is the synaptic coefficient from the j-th component of spatial units to the i-th hidden neuron, b_i is the bias value and s is the logistic activation function. Nelder-Mead simplex method is used for minimization problem and the given approximate solution works well for the points inside and outside the problem domain, near boundary points. Some advantages of this approach are that, it can solve the time dependent systems of partial differential equations, the method is generalized for solving the higher order and nonlinear problems, it deals with a few number of parameters, solution is fast evaluated, the method can be applied to initial and two point boundary value problem for ordinary differential equations, and it uses parallel processing. Unlike the other methods, there is no ill conditioning of the concluded linear system in the expansion methods or the necessity of making a special relation among the step size for different axis in the finite difference method.

In article Tsoulos et al. [46], used a hybrid method utilizing constructed feed-forward neural networks by grammatical evolution and a local optimization procedure, in order to solve ordinary differential equations, system of ordinary differential equations and partial differential equations.

4.1.5.9 MLP with Grammatical Evolution

Grammatical evolution is an evolutionary technique that can produce code in any programming language requiring the grammar of the target language in BNF syntax and some proper fitness function. The construction of neural network with grammatical evolution was introduced by Tsoulos et al. [47]. The method is based on an evolutionary algorithm whose basis is lies on the biological evolution and the efficiency of neural network is used as the fitness of the evolutionary algorithm along with a penalty function which is used in order to represent the initial or boundary value problem. The algorithm of constructed neural network for solving ordinary differential equation can be given as:

(i) Equidistant points are chosen within the interval $[a, b]$ and is denoted by $[x_1, x_2, \ldots, x_n]$.
(ii) The neural network $N(x, g)$ is constructed using grammatical evolution.
(iii) Training error has been calculated using the error function:

$$E(N(g)) = \sum_{i=0}^{T-1} \left(f(x_i, N(x_i, g), N^{(1)}(x_i, g), \ldots, N^{(n)}(x_i, g))\right)^2 \tag{4.49}$$

(iv) Penalty value $P(N(g))$ is calculated using the following equation:

$$P(N(g)) = \lambda \sum_{k=1}^{n} \psi_k^2(x, N(x, g), N^{(1)}(x, g), \ldots, N^{(n-1)}(x, g))_{|x=t_k} \tag{4.50}$$

where λ is a positive number.

(v) And finally, the fitness value is calculated as:

$$V(g) = E(N(x,g)) + P(N(x,g)) \tag{4.51}$$

The main advantage of the proposed method is that it has very less execution time and does not require a user to enter any parameter. This method can be easily parallelized, since it is based on the genetic algorithms and can be extended by using different BNF grammars for the constructed neural networks with different topologies or different activation functions. The proposed method does not require the user to enter any information regarding the topology of the network and the advantage of using an evolutionary algorithm is that the penalty function can be incorporated easily into the training process.

In [48], the authors attempted to present a novel method for solving fuzzy differential equations using multilayer perceptron neural network technique.

4.1.5.10 MLP for Fuzzy Differential Equations

Keeping in mind the function approximation capabilities of neural network authors presented a neural network model for solving fuzzy differential equations. A first order fuzzy differential equation has the form:

$$\chi'(t) = f(t, \chi(t)) \tag{4.52}$$

where χ is a fuzzy function of t and $f(t,\chi)$ is a fuzzy function of the crisp variable t and the fuzzy variable χ and χ' is the fuzzy derivative of χ, together with the initial condition $\chi(t_0) = \chi_0$. Equation (4.52) together with the boundary conditions is replaced by the following equivalent system:

$$\begin{cases} \bar{\chi}'(t) = \bar{f}(t,\chi) = F(t,\bar{\chi},\tilde{\chi}), & \bar{\chi}(t_0) = \bar{\chi}_0 \\ \tilde{\chi}'(t) = \tilde{f}(t,\chi) = G(t,\bar{\chi},\tilde{\chi}), & \tilde{\chi}(t_0) = \tilde{\chi}_0 \end{cases} \tag{4.53}$$

where $F(t,\bar{\chi},\tilde{\chi})$ and $G(t,\bar{\chi},\tilde{\chi})$ represents the minimum and maximum values of the function respectively and the equation is represented in the parametric form. The trial solution of the problem is written as the sum of the two parts in which first term satisfies initial or boundary condition and second term represents feed forward neural network. If $\bar{\chi}_T(t,r,\bar{p})$ is a trial solution for the first equation in system, as in Eq. (4.53), and $\tilde{\chi}_T(t,r,\tilde{p})$ is a trial solution for the second equation in Eq. (4.53) where $\bar{p}$ and $\tilde{p}$ are adjustable parameters, then the problem is transformed into the optimization problem. Each trial solution $\bar{\chi}_T$ and $\tilde{\chi}_T$ represents one feedforward neural network for which the corresponding networks are denoted by $\bar{N}$ and $\tilde{N}$ with adjustable parameters. Network is trained to satisfy the differential equations and the trial solutions should satisfy the differential equation. Quasi Newton method is used for the minimization problem and optimal values, weights are obtained after the optimization process. The neural network method for solving fuzzy differential equation has the following advantage:

(i) The approximated solution for fuzzy differential equation is very closed to the exact solution since neural networks have good approximation capabilities.
(ii) Solution of fuzzy differential equation is available at each training point within the interval.

In [48], authors presented a method for solving boundary value problem using artificial neural networks for irregular domain boundaries with mixed boundary conditions and introduce the concept of length factor for constructing approximate solution.

4.1.5.11 Development of Approximate Solution Using Length Factor

As defined above for the case of MLP the trial solution for boundaries with only Dirichlet and Neumann condition is defined respectively by

$$\psi_t = A_D(x) + F(x,N) \tag{4.54}$$

$$\psi_t = A_D(x) + A_M(x,N) + F(x,N) \tag{4.55}$$

where the first term A_D satisfy the Dirichlet and Neumann boundary condition in the equations respectively, A_M ensures the satisfaction of boundary condition while not interfering with the Dirichlet condition in Eq. (4.55) and the second term has to return zero on the boundary while being the function of the ANN output N for all the points inside the domain in Eq. (4.54). The concept of length factor is presented to produce the term for complex boundaries:

$$F(x,N) = N\, L_D(x) \tag{4.56}$$

and, A_M is defined as in Eq. (4.57)

$$A_M = L_D\, L_M\, g_M(x,N) \tag{4.57}$$

where the length factor L_D is a measure of distance from the Dirichlet boundary and L_M corresponds to the Neumann condition, g_M compensates the contribution of partial derivatives of A_D and F to obtain the desired boundary condition. Thin plates splines [49] are used to compute the length factor for the boundary conditions. The ANN output N is optimized to approximate solution in Eq. (4.55) and satisfies the differential equation in the domain as closely as possible. The method presented by the author is simpler than the other neural network method for irregular boundaries due to its unconstrained nature and automatic satisfaction of boundary conditions.

The continuous approximate solution can be easily evaluated at any point within the domain with higher accuracy.

Alli et al. in [50] presented an artificial neural network technique for the solution of mathematical model of dynamic system represented by ordinary differential equations and partial differential equations.

4.1.5.12 MLP for Vibration Control Problem

Because of non-linearity and complex boundary conditions, numerical solutions of vibration control problem always have some drawbacks such as numerical instability. That is why, an alternative method using feed forward artificial neural networks is presented by the authors for dynamical system. Extended backpropagation algorithm is used to minimize the objective function and is used for training ANN. The most commonly used objective function is taken, which is defined as:

$$\varepsilon^2 = |T_q - \phi_q|^2 \tag{4.58}$$

where, T_q is the target output and ϕ_q is the network output. In the case of extended backpropagation algorithm the weights are updated for the output neurons according to:

$$v(N+1) = v(N) - \eta_p(-2)|T_q - \phi_q|\phi_p \tag{4.59}$$

$$w(N+1) = w(N) - \eta_p(2)|T_q - \phi_q|v\alpha\phi_p|1 - \phi_p| \tag{4.60}$$

$$u(N+1) = u(N) - \eta_p(2)|T_q - \phi_q|v\alpha\phi_p|1 - \phi_p| \tag{4.61}$$

where u, v, w are weight parameters, η represents the learning rate, hidden and output layer are indexed as p and q respectively. Then the method is applied to many controlled and non controlled vibration problems of flexible structures whose dynamics are represented by ODE's and PDE's, for e.g., they considered mass-damper-spring system, whose mathematical model is given by:

$$m\frac{d^2\psi}{dt^2} + c\frac{d\psi}{dt} + k\psi = 0 \tag{4.62}$$

where the initial conditions of the systems are:

$$\psi(0) = 1 \quad \text{and} \quad \frac{d\psi(0)}{dt} = 0 \quad \text{with} \quad t \in [0, 2]$$

The authors also considered the second and fourth order PDEs that are the mathematical models of the control of longitudinal vibrations of rods and lateral vibration of beams. To test their method, they also obtain the solutions of the same problems by using analytical and Runge-Kutta method. It has been also observed that the presented method also success outside the training points when the neuron numbers in the hidden layer are increased.

An algorithm for the selection of both input variables and a sparse connectivity of the lower layer of connections in feed forward neural network of multilayer perceptron with one layer of hidden non linear single linear output node is presented by Saxen and Pettersson in [51].

4.1.5.13 Method for Selection of Inputs and Structure of FFNN

The algorithm for the selection for the inputs and structure of feed forward neural network of MLP type with a single layer of hidden nonlinear unit and a single linear output node can be mentioned as:

(i) A set of A potential inputs x and the output y is estimated for the M observation of the training set.
(ii) Iteration index is set to $k = 1$, a sufficient number of hidden nodes are taken and the weights are randomly generated for the lower layer of connections of the network.
(iii) Each non zero weight in weight matrix is turns zero and determine the optimal upper layer weights. Corresponding value of the objective function $F_{ij}^{(k)}$ is saved.
(iv) Find the minimum of the objective function values and set $W^{(k)} = W^{(k-1)}$ and equate to zero the weight corresponding to the minimum objective function value.
(v) Set $\psi_{ij} = k$ and save this variable in book keeping matrix then set $K = K + 1$, if $k < mn$ the algorithm repeats and go to (II) otherwise stops.

The results of the algorithm are saved in a book-keeping matrix that can be interpreted in retrospect to suggest a suitable set of inputs and connectivity of the lower part of the network. Various test examples are presented by the authors to illustrate that the proposed algorithm is a valuable tool for the users in extracting relevant inputs from a set of potential ones. The method is a systematic method that can guide the selection of both input variables and sparse connectivity of the lower layer of connections in feed forward neural networks of multi-layer perceptron type with one layer of hidden nonlinear units and a single linear output node and the algorithm developed for the method is efficient, rapid and robust.

Filici in [52], presented a method of error estimation for the neural approximation of the solution of an ordinary differential equation.

4.1.5.14 Error Estimation in Neural Network Solution

The author adopted the ideas presented by Zadunaisky [53, 54] in order to provide a method that can estimate the errors in the solution of an ordinary differential equation by means of a neural network approximation. Firstly neural approximation to the ordinary differential equation problem is computed, and then neural neighboring problem is solved and the true error $\bar{e}$ is estimated. A bound on the difference between the errors e and their estimations $\bar{e}$ is derived, which is used to provide an heuristic criterion for the validity of the error estimation under some assumptions. Let $e(t)$ be the solution of true error and $\bar{e}(t)$ is the solution of approximate error, it is assumed that for $\alpha \in [-1, 1]$, $N(t) + \alpha x(t)$ belong to S for all $t \in [0, t_f], x \in B_\Gamma$, it is also assumed that $\theta_i(t)$ and $\bar{\theta}_i(t)$ are continuous in t $\forall i$. Then $\exists$ positive constants γ and μ such that

$$||e - \bar{e}|| \leq \gamma \exp Lt + \frac{\mu}{L}[\exp(Lt) - 1] \tag{4.63}$$

$\forall\, t \in [0, t_f]$, with $||\cdot||$ is the Euclidean norm. A set of examples are presented by the author to show that the method can provide reasonable estimation of true errors. These examples also show that the criterion of validity works well in assessing the performance of the method.

Dua in [55], proposed a new method for parameter estimation of ordinary differential equations, which is based upon decomposing the problem into two sub problems.

4.1.5.15 MLP Based Approach for Parameter Estimation of System of ODE

The first sub problem generates an artificial neural network model from the given data and then the second sub problem uses the artificial neural network model to obtain an estimate of the parameters of the original ordinary differential equation model. The analytical derivatives from the artificial neural network model obtained from the first sub problem are used for obtaining the differential terms in the formulation of the second sub problem. The author considered a problem $P1$ as:

$$\varepsilon_1 = \min_{\theta, z(t)} \sum_{i \in I} \sum_{j \in J} (\hat{z}_j(t_i) - z_j(t_i))^2 \tag{4.64}$$

subject to:

$$\frac{dz_j(t)}{dt} = f_j(z(t), \theta, t)\; j \in J, \quad z_j(t = 0) = z_o\; j \in J, \quad t \in [t_o, t_f]$$

and constructed the sub problem $P2$ and $P3$ of $P1$. Problem $P3$ involves only algebraic variables, θ, and therefore it can be solved to global optimality more efficiently than the original problem $P1$ involving differential as well as algebraic variables, z and θ. The author recognized that a simpler sub problem $P3$ is obtained by solving the first sub problem, to obtain the artificial neural network model. The proposed approach is tested on the various example problems and encouraging results have been obtained. The main advantage of the proposed method is that requirement of high computational resources is avoided for computing the solution of a high optimization problem and instead of that two sub problems are solved. This approach is particularly useful for large and noisy data sets and nonlinear models where artificial neural networks are known to perform well.

In article [56], the authors considered optimal control problems of discrete nonlinear systems. Neural network is used to train the find function value, solution of the Hamilton-Jacobi-Bellman equation and the optimal control law.

4.1.5.16 MLP with Two-Coupled Neural Network

They considered an invariant system; that is function f is independent from the time k. Moreover, the cost increment function r is also considered as independent of the time k. Taking horizon N as infinite, they deduced a equation:

$$\frac{\partial r}{\partial u} + \left(\frac{\partial f}{\partial u}\right)^T \cdot \frac{\partial I}{\partial x}[f(x,u)] = 0 \tag{4.65}$$

Equation (4.65) is difficult to solve and analytical solutions are not usually possible to obtain, due to the nonlinearity of the problem. Thus, the author used neural networks to solve the problem based on the intelligent method such as neural network. They proposed to use two coupled neural networks to solve the Hamilton-Jacobi-Bellman equation in order to approximate nonlinear feedback optimal control solutions. The first neural network computes the value function $I(x)$ and the second one determines the optimal control law $g(x)$. The learning of two neural networks can be described as in Fig. 4.3.

The value function $I(x)$ corresponds to the output of the first neural network as:

$$I(x) = h_1(\beta) \tag{4.66}$$

where, $\beta = \sum_j W_{ji}^2 s_j + b_1^2$ and the output of the second neural network is:

$$u_m(x) = h_2(\beta_{u,m}) \tag{4.67}$$

where $\beta_{u,m} = \sum_j W_{u,jm}^2 s_{u,j} + b_{u,m}^2$, h_2 is the activation function of the output layer.

Minimization has been done and the weights are updated using gradient descent rule. In order to test the robustness of the proposed method, they consider two kinds of perturbations. The first type is obtained by variations on the parameters of the

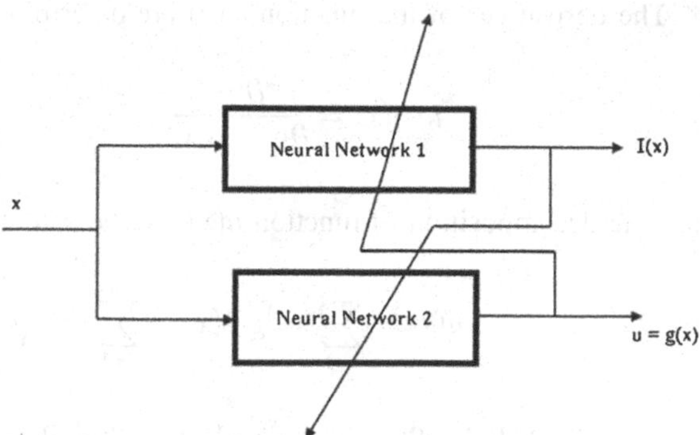

Fig. 4.3 Structure of two coupled neural network

system and in the second type of perturbation, they considered random noises caused by that sensor imperfections which affect the measured state variables. Simulation results show the performance and robustness of the proposed approach.

4.2 Method of Radial Basis Function Neural Networks

Another method to find an approximate particular solution of a differential equation is achieved by using Radial Basis Functions (RBFs) as described in [57]. The advantage of Radial Basis function is that a single independent variable is involved regardless of the dimension of the problem. RBFs are particularly attractive when the domain can't be expressed as product domains of lower dimensions. The method discussed in previous section presents mesh-free procedures for solving linear differential equations (ODEs and PDEs) based on multiquadric radial basis function networks (RBFNs) and the simulation results are perfect. But it doesn't determine the number of basis functions, centres and widths of the RBFs. Many kinds of methods are proposed to optimize the complexity of RBFs in the existing literature [58, 59].

RBF neural network method to solve differential equations relies on the whole domain and the whole boundary instead of the data set and can obtain all the parameters at the same time. The function $u(x)$ to be approximated is defined by $u(x) : R^p \rightarrow R^1$ and decomposed into basis functions as

$$u(x) = \sum_{i=1}^{m} w^{(i)} g^{(i)}(x) \tag{4.68}$$

where the parameters, $w^{(i)}, c^{(i)}, a^{(i)}, b^{(i)},\ i = 1, 2, \ldots, m$, are to be obtained.

The derivatives of the function $u\,(x)$ are determined as:

$$u_{j\ldots l}(x) = \frac{\partial^k u}{\partial x_j \ldots \partial x_l} = \sum_{i=1}^{m} w^{(i)} \frac{\partial^k g^{(i)}}{\partial x_j \ldots \partial x_l} \tag{4.69}$$

and the decomposition of function $u(x)$ can be written as:

$$u(x) = \sum_{i=1}^{m} w^{(i)} g^{(i)}(x) = \sum_{i=1}^{m} w^{(i)} \sqrt{(r^2 + a^{(i)^2})} \tag{4.70}$$

To explain the radial basis function neural network method for solving the differential equations, 2D Poisson's equation over the domain Ω is considered

$$\nabla^2 u = p(x), \quad x \in \Gamma \tag{4.71}$$

subject to Dirichlet and/or Neumann boundary conditions over the boundary Γ

$$u = p_1(x), \quad x \in \Gamma_1 \tag{4.72}$$

$$n \cdot \nabla u = p_2 x \quad x \in \Gamma_2 \tag{4.73}$$

where ∇^2 is the Laplace operator, x is the spatial position, p is a known function of x and u is the unknown function of x to be determined; n is the outward unit normal; ∇ is the gradient operator; $\partial\Gamma_1$ and $\partial\Gamma_2$ are the boundaries of domain such as $\partial\Gamma_1 \cup \partial\Gamma_2 = \partial\Gamma$ and $\partial\Gamma_1 \cap \partial\Gamma_2 = \phi$; p_1 and p_2 are known functions of x.

The solution u and its derivatives can be expressed in terms of basis functions given in Eq. (4.70). The design of network is based on the information provided by the given differential equation and its boundary conditions.

In this method, the model u being decomposed into m basis functions in a given family as represented by Eq. (4.70) and the unknown parameters $w^{(i)}, c^{(i)}, b^{(i)}, i = 1, 2, \ldots, m$ are to be obtained by minimizing the following integration

$$e = F(w, c, a) = \int\!\!\int_{\Omega} [u_{11}(x) + u_{22}(x) - p(x)]^2 \, dx_1 dx_2 + \int_{\partial\Omega} [u(x) - p_1(x)]^2 ds + \int_{\partial\Omega} [n_1 u_1(x) + n_2 u_2(x) - p_2(x)]^2 ds \tag{4.74}$$

where

$$w = \left(w^{(1)}, w^{(2)}, \ldots, w^{(m)}\right), \quad c = \left(c^{(1)}, c^{(2)}, \ldots, c^{(m)}\right), \quad a = \left(a^{(1)}, a^{(2)}, \ldots, a^{(m)}\right)$$

That is to solve the following equation series

$$\frac{\partial F(w,c,a)}{\partial w^{(i)}}=0,\quad \frac{\partial F(w,c,a)}{\partial c^{(i)}}=0,\quad \frac{\partial F(w,c,a)}{\partial a^{(i)}}=0,\quad i=1,2,\ldots,m \tag{4.75}$$

Finally, the solution is to obtain the values of the parameters by Eq. (4.75).

Remark 2 It has been observed that radial basis function neural network method provides more accurate and attractive results in comparison to multiquadric radial basis function. Also RBFN method is suitable for solving ODE as well as PDE's problems and the conditions on boundary are not necessarily strict.

4.3 Method of Multiquadric Radial Basis Function Neural Network

It has been already proved in [60] that radial basis function networks (RBFNs) with one hidden layer are capable of universal approximation. For problems of interpolation and approximation of scattered data, there is a body of evidence to indicate that the multiquadric (MQ) function yields more accurate results in comparison with other radial basis functions [7, 61]. Mai-Duy and Tran-Cong [62, 63] have developed a new method based on the RBFNs for the approximation of both functions and their first order higher derivatives and named as direct radial basis function networks (DRBFN) and indirect radial basis function networks (IRBFN) procedure; it was also found that the IRBFN method yields consistently better results for both functions and its derivatives.

4.3.1 DRBFN Procedure for Solving Differential Equations

To explain the solution of differential equations using DRBFN and IRBFN procedures Mai-Duy and Tran-Cong considered in [63] the 2D Poisson's equation over the domain Ω

$$\nabla^2 u = p(x),\quad x\in\Omega \tag{4.76}$$

where ∇^2 is the laplacian operator, x is the spatial position, p is a known function of x and u is the unknown function of x to be found. Equation (4.76) is subject to Dirichlet and/or Neumann boundary conditions over the boundary Γ.

$$u = p_1(x),\quad x\in\Gamma_1 \tag{4.77}$$

$$n \cdot \nabla u = p_2 x \quad x \in \Gamma_2 \tag{4.78}$$

where n is the outward unit normal; ∇ is the gradient operator; Γ_1 and Γ_2 are the boundaries of domain such as $\Gamma_1 \cup \Gamma_2 = \Gamma$ and $\Gamma_1 \cap \Gamma_2 = \phi$; p_1 and p_2 are known functions of x.

Since numerical solution of differential equation is intimately connected with approximating function and its derivatives. So the solution u and its derivatives can be approximated in terms of basis function. The design of neural network is based on the information provided by the given differential equation and its boundary conditions.

In the direct approach the sum squared error associated with Eqs. (4.76)–(4.78) is given by

$$SSE = \sum_{x^{(i)} \in \Omega} \left[(u_{,11}) x^{i} + u_{,22} \left(x^{(i)}\right) - p\left(x^{(i)}\right) \right]^2 + \sum_{x^{(i)} \in \Gamma_1} \left[u\left(x^{(i)}\right) - p_1\left(x^{(i)}\right) \right]^2 + \sum_{x^{(i)} \in \Gamma_2} \left[n_1 u_{,1}\left(x^{(i)}\right) + n_2 u_{,2}(x^{(i)}) - p_2(x^{(i)}) \right]^2 \tag{4.79}$$

A system of linear algebraic equation is obtained in terms of known weights in the output layer of the network by putting expression for u and its derivatives which have already been calculated in [64] in Eq. (4.79), as follows:

$$(G^T G) w = G^T \hat{p} \tag{4.80}$$

where G is the design matrix whose rows contains basis functions corresponding to the terms $(u_{,11}(\mathrm{x}^{(i)})) + (u_{,22}(\mathrm{x}^{(i)})), u(\mathrm{x}^{(i)})$ and $(n_1 u_{,1}(\mathrm{x}^{(i)}) + n_2 u_{,2}(\mathrm{x}^{(i)}))$ and therefore the number of rows is greater than the number of columns (number of neurons); w is the vector of weights and $\hat{p}$ is the vector whose elements correspond to the terms $p\ (\mathrm{x}^{(i)})$, $p_1(\mathrm{x}^{(i)})$, $p_2\ (\mathrm{x}^{(i)})$.

The solution u in the least squares sense in Eq. (4.79) can be obtained by using the method of orthogonal triangular decomposition with pivoting or the QR method [65] for an over determined system of equations, which is

$$Gw = \hat{p} \tag{4.81}$$

In practice, the QR method is able to produce the solution at larger values of the width of the basis function, than the normal equations method arising from the linear least square procedures Eq. (4.79).

4.3.2 IRBFN Procedure for Solving Differential Equations

In indirect method of approximation, in [63] the function u is obtained via a particular $u_{,jj}$ which is generally only one of a number of possible starting points. For the method to be correct, all starting points must lead to same value for function u. Thus, in indirect approach all possible starting points are taken into account and the sum squared error is given by

$$\begin{aligned} SSE &= \sum_{x^{(i)}\in\Omega} [(u_{,11}(x^{(i)}) + u_{,22}(x^{(i)}) - p(x^{(i)}))] \\ &+ \sum_{x^{(i)}\in\Omega} [u_1(x^{(i)}) - u_2(x^{(i)})]^2 + \sum_{x^{(i)}\in\Gamma_1} [u_1(x^{(i)}) - p_1(x^{(i)})]^2 \\ &+ \sum_{x^{(i)}\in\Gamma_2} [n_1 u_{,1}(x^{(i)}) + n_2 u_2(x^{(i)}) - p_2(x^{(i)})]^2 \end{aligned} \tag{4.82}$$

where the term $u_1(x^{(i)})$ is obtained via $u_{,11}$ and $u_2(x^{(i)})$ is obtained via $u_{,22}$. Furthermore, the unknown in the direct procedure also contains the set of weights introduced by the interpolation of the constants of integration in the remaining independent co-ordinate directions.

Remark 3 The DRBFN method yields similar accuracy to other existing methods [66–68] etc. On the other hand, the IRBFN method produces results which are several orders of magnitude more accurate than those associated with the DRBFN method if accuracy is measured in terms of norm of error.

The ease of preparation of input data, robustness of DRBFN and IRBFN method and high accuracy of the solution make the methods very attractive in comparison with conventional methods such as the FDM, FEM, FVM and BEM.

The indirect RBFN procedure achieves better accuracy than the direct RBFN procedure over a wide range of width of basis function and hence the choice of RBF width is less critical in the case of IRBFN procedure. Unlike the MLP neural network method, RBFN method is not iterative and hence more efficient. Both regularly shaped and irregularly shaped domains can be handled with this method.

4.3.3 Recent Development in the RBF and MQRBF Neural Network Techniques

In [69], the authors presented the combination of new mesh free radial basis function network (RBFN) methods and domain decomposition (DD) technique for approximating functions and solving Poisson's equation.

4.3.3.1 Radial Basis Function Domain-Decomposition Approach

Since IRBFN procedure achieves greater accuracy than DRBFN over a wide range of RBF widths for function approximation [62], therefore the IRBFN method is considered in conjunction with a domain decomposition technique for approximation of function and solving partial differential equations particularly Poisson's equation. In the IRBFN technique described previously, for approximation of the function of several variables and its derivative each derivative f_j and the associated function f_j is represented by an IRBFN and trained independently for small system of equations. A better approach should be that both sets will give the same approximation $f_1 = f_2$, hence w_1 and w_2 are solved simultaneously with the consequence that the system of equation is larger. They developed a new feature of IRBFN method for the approximation, so that the difficulties related to solving big matrices can be overcome by using a subregioning technique. Each sub region is approximated by a separate RBFN and the network is trained independently and, if desired in parallel. Subregioning of the domain provides an effective means of keeping the size of the system matrices down while improving accuracy with increasing data density. Authors developed the boundary integral equations based domain decomposition method for the estimation of boundary conditions at interfaces in solving given Poisson's equation of potential problem.

$$\nabla^2 u = b, \quad x \in \Omega \tag{4.83}$$

$$u = \bar{u}, \quad x \in \delta\Omega_u \tag{4.84}$$

$$q = \frac{\partial u}{\partial n} = \bar{q}, \quad x \in \partial\Omega_q \tag{4.85}$$

where u is potential, q is the flux across the surface with unit normal, $n, \bar{u}$ and $\bar{q}$ are the known boundary conditions, b is known function of position and $\partial\Omega = \partial\Omega_u + \partial\Omega_q$ is the boundary of the domain Ω. In their method, the interface boundary conditions are first estimated by using boundary integral equations (BIEs) at each iteration and sub domain problems are then solved by using RBFN method. Also the volume integrals in standard integral equation representation (IE), which usually require volume discretisation, are completely eliminated in order to present the mesh free RBFN method. The convergence rate of the approach can be affected by the element type used to compute BIEs. The numerical examples show that the RBFN methods in conjunction with domain decomposition technique not only achieve a reduction of memory requirement but also a high accuracy of the solution. The boundary integral equation based domain decomposition method is very suitable for coarse-grained parallel processing and can be extended to those problems whose governing equation can be expressed in terms of integral equations such as viscous flow problems.

Jianyu et al. in [70] defined a neural network for solving PDE in which activation function of the hidden nodes are the RBF and whose parameters are determined by the two stage gradient descent strategy.

4.3.3.2 Two-Stage Gradient Learning Algorithm for RBF

The authors illustrated the two stage gradient learning algorithm by considering 2D Poisson's equation

$$\Delta u = P(x), \quad x \in \Omega \tag{4.86}$$

where Δ is Laplace operator, x is the spatial function, P is known function of x and u is the unknown function of x to be found subject to the Dirichlet and Neumann boundary conditions over boundary

$$\begin{aligned} u &= P_1(x), & x \in \delta\Omega \\ n \times \nabla u &= P_2(x), & x \in \delta\Omega_2 \end{aligned} \tag{4.87}$$

n is the outward unit normal, ∇ is gradient operator, $\delta\Omega_1, \delta\Omega_2$ is the boundary of domain such that $\delta\Omega_1 \cup \delta\Omega_2 = \delta\Omega$ & $\delta\Omega_1 \cap \delta\Omega_2 = \phi$ and P_1, P_2 are known functions of x. They introduced a new incremental algorithm for growing RBF networks and a two stage learning strategy for training the network parameters. Now the model is decomposed into the set of m basis functions, the unknown parameters are obtained by minimizing the sum of square due to error. So the gradient descent optimization technique is used which works in two stages:

(I) $c^{(i)}$ and $a^{(i)}$ are fixed and $w^{(i)}$ is calculated by minimize the SSE by following formula:

$$w_t^{(i)} = w_{t-1}^{(i)} - \eta_{t-1} \frac{\partial l(c_{t-1}^{(i)}, a_{t-1}^{(i)}, w_{t-1}^{(i)})}{\partial w_{t-1}^{(i)}} \tag{4.88}$$

(II) $w_t^{(i)}$ is now fixed and $c^{(i)}$, $a^{(i)}$ is computed by minimizing the SSE by following equation:

$$c_i^{(t)} = c_{t-1}^{(i)} - \beta_{t-1} \frac{\partial l(c_{t-1}^{(i)}, a_{t-1}^{(i)}, w_t^{(i)})}{\partial c_{t-1}^{(i)}} \tag{4.89}$$

$$a_i^{(t)} = a_{t-1}^{(i)} - \alpha_{t-1} \frac{\partial l(c_t^{(i)}, a_{t-1}^{(i)}, w_t^{(i)})}{\partial a_{t-1}^{(i)}} \tag{4.90}$$

where η_{t-1}, β_{t-1} and α_{t-1} are the learning rates at time $t-1$ and can be decide by the recurrent procedure. This learning strategy is able to solve computational time and memory space because of the selective growing of nodes whose activation functions consists of different RBFs.

Kansa et al. [71] proposed a finite volume analog of the meshless RBF method for the solution of system of non linear time dependent partial differential equations.

4.3.3.3 Volumetric Integral Radial Basis Function Method

The integration approach presented by the authors is physically meaningful since only the extensive volume integration of the various density functions obey strict conservation laws. Integral form of conservation law can be given as:

$$\int_{\Omega}\left[\frac{\partial u}{\partial t}+\nabla \cdot F\right] dV = 0 \qquad (4.91)$$

The basis functions are modifies to be integrals of RBFs evaluated at the discrete knots which yield coefficient matrices that premultiply the column vectors. Physical domain decomposition over piecewise continuous sub domains are applied that are bounded by shocks, contact surfaces, or rarefaction fans. The authors converted the set of non linear multidimensional partial differential equations into a set of ordinary differential equations, by a series of rotational and translational transformations and introduce an additional local transformation that maps these ordinary differential equations into compatibility or eigen vector ordinary differential equations that propagate at 'characteristic' velocities, thereby decoupling the compatibility ordinary differential equations. By writing the compatibility variables as a series expansion of RBF's with time dependent expansion coefficients, the time advanced problem is converted into a method of lines problem. The volume integration of the RBF's is performed in parallel by integrating over each sub domain Ω_i separately, and normalizing the results. They tracked strong shocks, capture weak shocks using artificial viscosity methods to dampen them and used Riemann solvers and shock polar method for shock wave interactions. When pairs of knots coincide, discontinuous surface at the coincidence loci is introduced. Since volumetric integral formulation of time dependent conservation equations increases the convergence rates of radial basis function approximates, therefore fewer number of knots are required to discretize the domain.

Zou et al. in [72] presented a new kind of RBF neural network method based on Fourier progression, by adopting the trigonometric function as basis function.

4.3.3.4 RBF with Trigonometric Function

They used $\hat{W}B(x)$ to approximate a unknown function $f(x)$ where,

$$\begin{aligned} \hat{W} &= [\hat{w}_1, \hat{w}_2, \ldots, \hat{w}_n, \hat{w}_{n+1}, \ldots, \hat{w}_{2n}, \hat{w}_{2n+1}] \\ B(x) &= [\sin x, \ldots, \sin nx, \cos x, \ldots, \cos nx, c] \end{aligned} \tag{4.92}$$

with the Fourier progression theory that every continuous function $f(x)$ can be expressed as follows,

$$f(x) = c + \sum_{n=1}^{\infty} a_n \sin nx + \sum_{n=1}^{\infty} b_n \cos nx$$

and constructed a neural network which is dense for continuous function space. They constructed a optimal weight matrix by assuming that, a function vector $h : \Omega \to R^P$, for any $\sigma > 0$, there always exist a function array $B : R^m \to R^l$ and an optimal weight matrix W^* such that $||h(x) - W^{*T}B(x)|| \leq \sigma, \quad \forall x \in \Omega$, where Ω is a tight set of R^m and $h(x) - W^{*T}B(x) = \Delta h(x), \tilde{W} = \hat{W} - W^*$, where $\hat{W} \in R^{l\times 3}$ is used to estimate value of W^*. To apply the neural network to a practical system, a class of non linear systems was considered by the authors. Then it is used in a class of high order system with all unknown control function matrices. The adaptive robust neural controller is designed by using back stepping method and effectiveness of the method is presented by simulation study. It has been pointed out that by adopting the trigonometric function as basis function, the input needs not to be force between -1 and 1, and there is no need to choose the centre of basis function.

In article [73], the author presented a meshless method based on the radial basis function networks for solving high order ordinary differential equations directly.

4.3.3.5 RBF for Higher-Order Differential Equations

Two unsymmetric RBF collocation schemes, named the usual direct approach based on a differentiation process and the proposed indirect approach based on an integration process, are developed to solve high order ordinary differential equations. They considered the following initial value problem governed by the following p-th order ordinary differential equation

$$y^{|p|} = F(x, y, y', \ldots, y^{|p-1|}) \tag{4.93}$$

with initial conditions $y(a) = \alpha_1, y'(a) = \alpha_2, \ldots, y^{|p-1|}(a) = \alpha_p$, where

$$a \leq x \leq b, \quad y^{(i)}(x) = \frac{d^i y(x)}{dx^i}$$

F is a known function and $\{\alpha_i\}_{i=1}^{p}$ is a set of prescribed conditions. Like other meshless numerical methods, the direct RBF collocation approach is based on the differential process to represent the solution. In the proposed RBF collocation approach, the closed forms representing the dependent variable and its derivatives are obtained through the integration process. In the case of solving high order ODEs, difficulties to deal with multiple boundary conditions are naturally overcome with integrating constants. Analytical and numerical techniques for obtaining new basis functions from RBF's are discussed. Among RBFs, multiquadrics are preferred for practical use. Numerical results show that the proposed indirect approach performs much better than the usual direct approach. High convergence rates and good accuracy are obtained with the proposed method using relatively low number of data points.

In [74], the authors presented a new indirect radial basis function collocation method for numerically solving bi-harmonic boundary value problem.

4.3.3.6 RBFNN Approach for Bi-harmonic BVP

Authors considered the bi-harmonic equation:

$$\frac{\partial^4 v}{\partial x_1^4}+2\frac{\partial^4 v}{\partial x_1^2 \partial x_2^2}+\frac{\partial^4 v}{\partial x_2^4}=F \tag{4.94}$$

in the rectangular domain Ω with F being a known function of x_1 and x_2, which can be reduced to a system of two coupled Poisson's equations

$$\frac{\partial^2 \mathrm{v}}{\partial \mathrm{x}_1^2}+\frac{\partial^2 \mathrm{v}}{\partial \mathrm{x}_2^2}=\mathrm{u},\ \mathrm{x}\in\Omega;\quad \frac{\partial^2 \mathrm{u}}{\partial \mathrm{x}_1^2}+\frac{\partial^2 \mathrm{u}}{\partial \mathrm{x}_2^2}=\mathrm{F},\ \mathrm{x}\in\Omega \tag{4.95}$$

since, in the case when boundary data are

$$\left\{\mathrm{v}=\mathrm{r}(\mathrm{x}),\frac{\partial^2 \mathrm{v}}{\partial \mathrm{n}^2}=\mathrm{s}(\mathrm{x}),\ \mathrm{x}\in\partial\Omega\right\}$$

The use of two Poisson's equation is preferred as each equation has its own boundary condition. In this research article [74], the authors described the indirect radial basis function networks and proposed a new technique of treating integrating constant for bi-harmonic problems, by eliminating integration constant point wise subject to the prescribed boundary conditions. It overcomes the problem of increasing size of conversion matrices caused by scattered points and provides an effective way to impose the multiple boundary conditions. Two types of boundary conditions

$$\left\{v,\frac{\partial^2 v/\partial n^2}{u}\right\}\quad\text{and}\quad\left\{v,\frac{\partial v}{\partial n}\right\}$$

are considered. The integration constants is excluded from the networks and employed directly to represent given boundary conditions. For each interior point, one can form a square set of k linear equations with k being the order of PDE's, from which the prescribed boundary conditions are incorporated into the system via integration constants. This is advancement in the indirect radial basis function collocation method for the case of discretizing the governing equation in a set of scattered data points. The proposed new point wise treatment in article [74] overcomes the problem of increasing size of conversion matrices, and provides an effective way to implement the multiple boundary conditions without the need to use fictitious points inside or outside the domains or to employ first order derivatives at grid points as unknowns. This method is truly a meshless method, which is relatively easy to implement as expression for integration constants are given explicitly and this represents a further advancement in the case of IRBFN for the case of discretizing the governing equations on a set of scattered data points.

Golbabai and Seifollahi in [75] implemented RBF neural network method for solving the linear-integro differential equations.

4.3.3.7 RBFNN for Linear-Integro Differential Equations

They proposed the approach by considering the following equation:

$$Dy(x) - \lambda \int_{\Gamma} k(x,t)y(t)dt = g(x), \quad \Gamma = [a,b] \tag{4.96}$$

with the supplementary conditions as follows:

$$\begin{cases} Dy(x) = y'(x) + A_1(x)y(x), \\ y(\alpha_1) = \gamma_1, \end{cases} \tag{4.97}$$

$$\begin{cases} Dy(x) = y''(x) + A_1(x)y'(x) + A_2(x)y(x), \\ y(\alpha_1) = \gamma_1, \quad y'(\alpha_2) = \gamma_2. \end{cases} \tag{4.98}$$

where D is the differential operator, λ, γ_1, and γ_2 are constants, $\alpha_1, \alpha_2 \in \Gamma, A_1, A_2, g$ and k are known functions and y is the unknown function to be determined. For illustrating the method they rewrite Eq. (3.34) in the following operator form

$$Dy - \lambda Ky = g, \tag{4.99}$$

where

$$(Ky)(x) = \int_{\Gamma} k(x,t)y(t)dt$$

and used the collocation method which assumes discretisation of the domain into a set of collocation data. They assumed an approximate solution $y_p(x)$ such that it satisfies the supplementary conditions and quasi-Newton Broyden–Fletcher–Goldfarb–Shanno (BFGS) method is used for training the RBF network. The authors also described an algorithm which is used in their experiment, the main attraction of their algorithm is that it starts with a single training data and with a single hidden layer neuron, then continues the training patterns one by one and allows the network to go. Various numerical examples are considered to demonstrate the proposed idea and method. Golbabai and Seifollahi also described radial basis function neural network method for solving the system of non linear integral equations in Eq. (4.35). The result obtained by this approach [75] proves that the RBF neural network with quasi-Newton BFGS technique as a learning algorithm provides a high accuracy of the solution. Also the approach is quite general and appears to be the best among approximation methods used in the literature. This method is recommended by the author to use in solving a wide class of integral equations because of its ease of implementation and high accuracy. Moreover, the reported accuracy can be improved further by increasing the number of training data and the number of hidden units in the RBF network to some extent.

The research article [77] introduced a variant of direct and indirect radial basis function networks for the numerical solution of Poisson's equation. In this method they initially described the DRBFN and IRBFN procedure described by Mai-Duy and Tran-Cong in [69] for the approximation of both functions and their first and higher order derivatives.

4.3.3.8 RBFNN for Poisson's Equation

The authors illustrated the method by considering a numerical example of two-dimensional Poisson's equation:

$$\nabla^2 u = \sin(\pi x_1)\sin(\pi x_2) \tag{4.100}$$

where, $0 \le x_1 \le 1$ and $0 \le x_2 \le 1$ with $u = 0$ on whole boundary points. They consider 20 points, 11 of those were boundary points and 9 were interior points, and used multiquadric radial basis function method which Deng et al. in [67] had used. Then they computed the approximate solution by converting the Cartesian coordinate into polar:

$$\nabla^2 u = \frac{\partial^2 u}{\partial r^2} + \frac{1}{r}\frac{\partial u}{\partial r} + \frac{1}{r^2}\frac{\partial^2 u}{\partial \theta^2} \tag{4.101}$$

They found that the approximated solution of this new method is better than both DRBFN and IRBFN method on the Cartesian ones. Further, they applied this method to the two dimensional Poisson's equation in the elliptical region and achieved better accuracy in the terms of root mean square error. In the above approach [77], it has been shown that transformation of Poisson's equation into the polar coordinate can achieve a better accuracy than the DRBFN and IRBFN methods on the cartesian ones. Also, the accuracy of the IRBFN method is influenced by the width parameter of the radial basis functions such that this parameter must be in the special range and as it increases the condition number increases too, but in this method, variations of the width parameter of a basis function do not influence at the accuracy of the numerical solution. Hence the condition number is small and the obtained system is stable.

Chen et al. in [78] proposed a method that develops a mesh free numerical method for approximating and solving PDEs, based on the integrated radial basis function networks (IRBFNs) with adaptive residual sub sampling training scheme. Integrated radial basis function network for approximating and solving PDEs is described initially. In this article, the authors adopted the residual sub sampling scheme suggested in [79] to train the IRBF network. In the training process, neurons are added and removed based on the residuals evaluated at a finer point set, and the shape parameter adjusting scheme is modified for suiting the IRBF neuron behavior which is different from DRBF network. They simply considered the shape parameters by multiplying the distances between two neighbour neurons with a fixed coefficient and multiquadric function is taken as the transfer function of the neurons. Adaptive process for determining the locations of neurons in integrated radial basis function networks for approximating a one dimensional function is described by the training procedure. During the training procedure, two neurons whose centres are end points which are always kept fixed. Numerical examples are conducted to show the effectiveness of the method. Since IRBFNs are capable to smooth the derivative errors for solving PDEs, therefore with the proposed adaptive procedure, IRBFNs require less neurons to attain the accuracy than DRBFN. Approximation based on smooth IRBFNs is highly effective in approximating smooth functions, even if the neuron sets are relatively coarse. The adaptive method applied for training in this article is an effective technique for dealing with the steep and corner feature of the PDEs solutions; and the IRBF networks contribute to improve the accuracy of solving PDEs. Hence a combination of IRBF and adaptive algorithm is a promising approach for mesh free solutions of PDEs. This method can easily be applied for solving higher dimension problems and time dependent nonlinear equations. A survey on MLP and RBF neural network methods for solving differential equations is also presented in [80].

4.4 Method of Cellular Neural Networks

The state-of-the art of the cellular neural networks (CNN) paradigm shows that it produces an attractive alternative solution to the conventional numerical computation method [81, 82]. It has been intensively shown that CNN is an analog computing paradigm which performs ultra-fast calculations and provides accurate results. In research article [83] the concept of analog computing based on the cellular network paradigm is used to solve complex non-linear and stiff differential equations. In this method, equations are mapped into a CNN array in order to facilitate templates calculation. Complex PDEs are transformed into ODE having structures and the transformation is achieved by applying the method of finite differences. This method is also based on the Taylor series expansion.

The concept of Cellular Neural Networks (CNN) was introduced by Chua and Yang [81]. CNN method for solving complex and Stiff ODE is given by following steps:

4.4.1 Principle for CNN Templates Findings

According to general theory in nonlinear dynamics based on the linearization of the vector field [84], complex and stiff ODEs can be described by a unique vector field in a bounded region of R^n, which is given as:

$$\frac{dx}{dt} = A(x)[x - F(x)] \tag{4.102}$$

where $A(x)$ is $n \times n$ matrix function of x, F being the mapping of R^n to itself. In this approach complex ODEs are transformed into the form described in Eq. (4.102) in order to make them solvable by the CNN paradigm, since it is well known that Eq. (4.102) can easily be mapped into the form of CNN model [81, 83].

Let us consider the case of a system consisting of three identical oscillators of the Rossler type coupled in a Master-Slave-Auxiliary configuration. The master (x_1, y_1, z_1) + slave (x_2, y_2, z_2) + auxiliary (x_3, y_3, z_3) system under investigation are modeled by the following differential equations:

$$\frac{dx_{1,2,3}}{dt} = -\omega_{1,2,3}y_{1,2,3} - z_{1,2,3} + \in_{1,2,3}\left(x_{2,1,1} + x_{3,3,2} - x_{1,2,3}\right) \tag{4.103}$$

$$\frac{dy_{1,2,3}}{dt} = \omega_{1,2,3}x_{1,2,3} + a_{1,2,3}y_{1,2,3} \tag{4.104}$$

$$\frac{dz_{1,2,3}}{dt} = f_{1,2,3} + z_{1,2,3}\left(x_{1,2,3} - U_{1,2,3}\right) \tag{4.105}$$

where ω_i are natural frequencies of the oscillators, ε_i are the elastic coupling coefficients and, a_i, f_i, u_i are the system parameters.

Let us take Eqs. (4.103)–(4.105), which are good prototypes of complex and stiff ODEs, then transform them into the form:

$$\frac{d}{dt}\begin{bmatrix} x_{1,2,3} \\ y_{1,2,3} \\ z_{1,2,3} \end{bmatrix} = \begin{bmatrix} -\varepsilon_{1,2,3} & -\omega_{1,2,3} & -1 \\ +\omega_{1,2,3} & +a_{1,2,3} & 0 \\ 0 & 0 & -U_{1,2,3} \end{bmatrix}\begin{bmatrix} x_{1,2,3} \\ y_{1,2,3} \\ z_{1,2,3} \end{bmatrix} + \begin{bmatrix} \varepsilon_{1,2,3}\left(x_{2,1,1}+\varepsilon_{3,3,2}\right) \\ 0 \\ f_{1,2,3}+x_{1,2,3}.z_{1,2,3} \end{bmatrix} \tag{4.106}$$

from Eq. (4.102) one can show the existence of fixed points through by Eq. (4.107).

$$\frac{d}{dt}\begin{bmatrix} x_{1,2,3} \\ y_{1,2,3} \\ z_{1,2,3} \end{bmatrix} = 0 \tag{4.107}$$

By Eq. (4.107) fixed points can evaluate as follows:
Master system fixed point

$$\hat{X}_1 = \begin{bmatrix} x_{01} \\ y_{01} \\ z_{01} \end{bmatrix} \tag{4.108}$$

Slave system fixed point

$$\hat{X}_2 = \begin{bmatrix} x_{02} \\ y_{02} \\ z_{02} \end{bmatrix} \tag{4.109}$$

Auxiliary system fixed point

$$\hat{X}_3 = \begin{bmatrix} x_{03} \\ y_{03} \\ z_{03} \end{bmatrix} \tag{4.110}$$

Now vector field is to linearize around fixed points and this linearization around a non-zero equilibrium fixed point provides the possibility of modifying the non-linear part of the coupled system without changing the qualitative dynamics of the system. This statement can be materialized by:

$$AX_{1,2,3} \quad \rightarrow \quad A\hat{X}_{1,2,3} \tag{4.111}$$

Therefore Eq. (4.102) can be considered to evaluate the linear part of the vector field at the fixed points. This linear part is represented by 3 × 3 matrices defined as follows:

$$A_{\text{master}} = \begin{bmatrix} a_{11} & a_{12} & a_{13} \\ a_{21} & a_{22} & a_{23} \\ a_{31} & a_{32} & a_{33} \end{bmatrix} \tag{4.112}$$

$$A_{\text{slave}} = \begin{bmatrix} b_{11} & b_{12} & b_{13} \\ b_{21} & b_{22} & b_{23} \\ b_{31} & b_{32} & b_{33} \end{bmatrix} \tag{4.113}$$

$$A_{\text{auxiliary}} = \begin{bmatrix} c_{11} & c_{12} & c_{13} \\ c_{21} & c_{22} & c_{23} \\ c_{31} & c_{32} & c_{33} \end{bmatrix} \tag{4.114}$$

from which the corresponding CNN templates are derived under precise values of the model in Eqs. (4.103)–(4.105).

4.4.2 Design of the Complete CNN Processor

Now we have to design a CNN computing platform to investigate the issues of synchronization in the master-slave-auxiliary system modeled by Eqs. (4.103)–(4.105). The efficiency of the calculations using CNN makes it a good candidate to perform computations in the cases of high stiffness and therefore this is an appropriate tool to tackle the difficulties faced by the classical numerical approach when dealing with the computation of the model in Eqs. (4.103)–(4.105). Using the structure of the basic CNN [80], design the complete CNN processor to solve the above model, and thus the results are obtained from the complete CNN processor.

Remark 4 The Cellular Neural Network method gives accurate results which are very close to results in the relevant literature [85–89] etc. The Computation based on CNN paradigm is advantageous, since it provides accurate and ultra-fast solutions of very complex ODEs and PDEs and performs real time computing.

4.4.3 Recent Development in the Cellular Neural Network Technique

Kozek and Roska [90] presented a cellular neural network for solving Navier-stokes equation which describes the viscous flow of incompressible fluids.

4.4.3.1 Double-Time Scale CNN Model

As an example they investigated poisson's equation in a 2D rectangular domain with some planar intensity function at a given location. In order to obtain the CNN model the spatial differential terms are substituted by central difference formulas and this discretization is don with equal step sizes in each direction. Hence an approximate expression for second order poisons equation can be obtain and using this approximation CNN template is designed. With the homogeneous term applied as a bias map a CNN array is obtained which, when started from a suitable initial condition and provided that the transient remains bounded, solves the poisson equation and the steady state of the CNN array gives the solution to the Poissons equation.

Authors constructed a three layer CNN model for Navier Stokes equation whose characteristic equation for incompressible fluids has the form:

$$\frac{\partial \vec{u}}{\partial t} + (\vec{u}\, grad)\vec{u} = \vec{f} - \frac{1}{e} grad\, p + v\nabla^2 \bar{u} \tag{4.115}$$

where $\vec{u}$ corresponds the velocity flow, p is pressure and $\vec{f}$ represents the effects of external forces. The Navier-stokes equation is converted into the 2D rectangular coordinates and expressed in the two conservative forms with pressure field is computed by Poisson equation. As the position equation these three equations are taken as the starting point in the CNN and the spatial derivatives are replaced by the difference terms. Since each CNN cell has only one output so three layers are constructed to represents the variables u, v and p and the CNN templates representing each variables are presented. Numerical simulation has been done for solving Navier stokes equation and the stationary flow pattern is shown for a source drain pair along with the corresponding pressure surface.

In [91] authors presented an analog cellular neural network method with variable mesh size for partial differential equations.

4.4.3.2 CNN with Variable Mesh Size

They introduce how the accuracy of the method can be improved by using a variable mesh size within the limited size of the neural network. A one dimensional problem has been considered as:

$$\frac{\partial^2 u}{\partial x^2} = f(x) \tag{4.116}$$

where $x_0 \le x \le x_{n+1}$, $u(x_0) = u_0$, $u(x_{n+1}) = u_{n+1}$. The second order partial derivative of the Eq. (4.116) is approximated by the difference equation with equal mesh size, then a system of linear equations is obtained in the matrix form

$$Au + \phi + b = 0 \tag{4.117}$$

where matrix A is symmetric and positive definite, then a convex energy function can be obtained as:

$$E(v) = \frac{1}{2} v^T A v + v^T \phi \tag{4.118}$$

A neural network is constructed to minimize the error function. For the problem considered above the resulting neural network is a row of cells and the cell is described by the following dynamic equation

$$\frac{du_i}{dt} = \frac{1}{CR_{i-1}} v_{i-1} - \left(\frac{1}{CR_{i-1}} + \frac{1}{CR_i} \right) u_i + \frac{1}{CR_i} v_{i+1} + \frac{I_i}{C} \tag{4.119}$$

At steady state $\frac{du_i}{dt} = 0$, and u_i represents the approximate solution of Eq. (4.116). They also find out that the mesh size is dependent on the resistance of the CNN, mesh size can be changed by simply changing the resistance R_i in the cellular neural network. Thus the technique has been developed to use variable mesh sizes, and thus to control the accuracy of the method for a particular number of neuron shells.

In [92] authors presented a cellular neural network technique for solving ordinary as well as partial differential equations.

4.4.3.3 CNN for Partial Differential Equations

They investigated the applicability of CNN to auto waves and spiral waves in reaction diffusion type system, Burgers equation and Navier-stokes equation for incompressible fluids. The solution obtained using CNN has the following four basic properties:

(a) It is continuous in time.
(b) Continuous and bounded in value.
(c) Continuous in interaction parameters.
(d) Discrete in space.

The non linear PDE

$$\frac{\partial u(x,t)}{\partial t} = \frac{1}{R} \frac{\partial^2 u(x,t)}{\partial x^2} - u(x,t) \frac{\partial u(x,t)}{\partial x} + F(x,t) \tag{4.120}$$

is considered which describes the mean density of moving particles along the coordinate x as a function of time t under the assumption that the particle velocity decreases linearly with the particle density. The spatial derivatives are replaced by the difference terms and Eq. (4.120) is approximated by the set of ordinary differential equations. Then compared the coefficient of the ODE to the state equation of a non linear CNN, hence the templates are calculated directly. Burgers equation have been solved for different values of R, Δx and for different initial condition $u(x,0)$. Authors presented the approximation accuracy of the CNN solution for homogeneous case by comparing them with the well known explicit solutions of Burgers equation and the results shows that the solution is strongly dependent on the parameter value of R, while for larger values of R two distinct peaks are observed that shifts to larger argument x as time increases. The examples given by the author in the development of the CNN architecture for solving a class of PDE's allow us to fully exploit the immense computing power offered by this programmable, analog, parallel computing structure.

In the previous paragraph authors in [92] presented the various techniques for converting various types of partial differential equations into equivalent cellular neural networks.

4.4.3.4 Reaction Diffusion CNN

Here Roska et al. in [93] presented the CNN solution of the equation of motion for chain of particles with non linear interactions, solution of non linear Klein-Gordon equation and application of a reaction diffusion CNN for finger print enhancement. One dimensional system of non linear differential equations has been considered

$$m\frac{d^2y_i}{dt^2} = -k\,[\,(y_i - y_{i-1}) - (y_{i+1} - y_i\,)\,] - c[\,(y_i - y_{i-1})^2 - (y_{i+1} - y_i)^2\,] \quad (4.121)$$

for $i = 1,\ldots, N - 1$ which represents the equation of motion for a non linear chain of particles. To apply the CNN approach the Eq. (4.121) is written in the set of ordinary differential equations which is equivalent to the two layer CNN with templates. The long term behavior of the motion of the chain has been examined and a considerable transmission of energy between the modes is obtained. Similarly the Klein-Gordon and reaction diffusion are solved using a CNN model and demonstrated that the CNN approach is a flexible framework for describing a wide variety of non linear phenomenon and also provides efficient implementation of such systems.

In [94] a learning procedure has been presented by the authors and applied it in order to find the parameters of the networks approximating the dynamics of certain nonlinear systems which are characterized by partial differential equations.

4.4.3.5 CNN for Nonlinear Partial Differential Equations

Partial differential equations can be solved by the CNN based on finite difference approximations and results in the set of ordinary differential equations. These set of differential equations can be represented by a single layer CNN with state equations. For learning the dynamics of a given non linear system, initially the basic CNN architecture has to be determined which includes the choice of neighborhood and of a class of non linear weight functions. While the number of cells in the neighborhood has to be large enough for estimating the highest order spatial derivatives in considered partial differential equation. Mean square error is prepared by assuming that the values of a special solution of a PDE for an initial condition $u^*(x,0)$ are known at the cell positions x_i for few times t_m as:

$$e(p_a) = \frac{1}{MN}\sum_{m=1}^{M}\sum_{i=1}^{N}\left[u_{CNN}(x_i,t_m,p_a) - u^*(x_i,t_m)\right]^2 \tag{4.122}$$

where p_a are the random initial values for the components of parameter vector and minimization is done by the simplex method. The results obtained by the method shows the dynamics of the non linear systems and more accurate than those using direct discretization.

A CNN method for solving a class of partial differential equation is presented in [95] where each neural cell consists of a novel floating variable, linear resistor, an amplifier and a passive capacitor.

4.4.3.6 CMOS VLSI Implementation for CNN

Each cell in neural network is integrated with a micro electromechanical heater device for the output. Initially it is assumed that the region is a rectangle containing the square grid of points $P_{i,j}$ spaced a distance h apart. The accuracy of the partial differential equation is increased by decreasing the mesh distance h but it increases the expenses with VLSI circuits. So mesh density is increased only in the areas where the function is most active. The system for a non equidistant mesh is defined by

$$N_{i,j} = \left[\frac{2}{n(n+s)}\right]_{i,j} \tag{4.123}$$

$$S_{i,j} = \left[\frac{2}{s(n+s)}\right]_{i,j} \tag{4.124}$$

$$E_{i,j} = \left[\frac{2}{e(e+w)}\right]_{i,j} \tag{4.125}$$

$$W_{i,j} = \left[\frac{2}{w(e+w)}\right]_{i,j} \tag{4.126}$$

$$\beta_{i,j} = N_{i,j} + S_{i,j} + E_{i,j} + W_{i,j} \tag{4.127}$$

where n is the distance above the node north, s is the south, e and w are the east and west respectively. Then an energy function has been prepared using the matrix form for the variable mesh case as:

$$E(v) = \frac{1}{2} v^T A v + v^T \phi \tag{4.128}$$

The layout for the entire nine cell circuit is presented and Matlab was used to obtain node values to compare with the simulation for the 3×3 array of CNN cells. The chip layout was analyzed and the results show that the circuit matches the simulations with a small amount of error.

In [96] authors presented a method to model and solve direct non linear problem of heat transfer in solids by using the cellular neural network.

4.4.3.7 CNN to Solve Nonlinear Problems of Steady-State Heat Transfer

Problem statement is defined by taking 2D steady state heat conduction in the observed solid having continually distributed internal heat sources as:

$$\lambda \left[\frac{\partial^2 T}{\partial x^2} + \frac{\partial^2 T}{\partial y^2} \right] + q_v = 0 \tag{4.129}$$

where T is temperature of solid, λ is thermal conductivity of the solid and q_v is the volumetric heat flow rate of internal heat sources with the boundary condition of first, second, third and fourth kind. Linearization of the equation has been done by using Kirchhoff's law, so after the transformation boundary condition of first and second kind becomes linear while the boundary condition of third and fourth kind remains non linear. Modeling and solution of Eq. (4.129) are performed by modified two-dimensional orthogonal CNN and each $M \times N$ node of CNN contains the multi input and multi output cell $C_{i,j}$. The problem is solved by two levels of the CNN: the lower one determined by the other feedbacks and the upper one determined by another feedback, hence the network is named as single layer two level CNN.

Authors in [97] proposed an implementation of a cellular neural network to solve linear second order parabolic partial differential equations.

4.4.3.8 CNN for Parabolic PDE

A two dimensional parabolic equation is considered which is defined on a region R as:

$$\frac{\partial u}{\partial t} = A \frac{\partial^2 u}{\partial x^2} + B \frac{\partial^2 u}{\partial y^2} + D \tag{4.130}$$

where $u(x, y, t)$ is a continuously unknown scalar function that satisfies given set of boundary conditions, A, B and D are functions of spatial dimension x and y. Equation (4.130) is written in the difference form of the derivatives to get the numerical solution. In order to solve general parabolic PDE's the resistors and capacitors in the circuit should be programmable and controlled by a digital input. 8×8 cell network was simulated with the specific circuit simulator to show the feasibility of the circuit simulator to solve parabolic partial differential equation. Two version of this circuit were taken in which in the one circuit, all the resistors and capacitors were ideal, passive elements for comparison purpose and another circuit had switched capacitor equivalent resistors, capacitor banks and local memories. To achieve zero Dirichlet boundary condition all the boundary nodes were connected to the ground, at first initial values were loaded into the local memories and then converted into analog voltages by the D/A converters. Circuit performance is also given by the simulation results which show that CNN technique succeeds in improving solution throughout and accuracy.

In [98] an online learning scheme to train a cellular neural network has been presented which can be used to model multidimensional systems whose dynamics are governed by partial differential equations.

4.4.3.9 Training CNN Using Backpropagation Algorithm

A CNN is trained by modified back propagation algorithm and the goal of the training is to minimize the error between the outputs of the trained CNN to the training data by means of adjusting the parameter values. Templates are unknown to this case:

$$\hat{u}_{ij} = -a\hat{u}_{ij} + P^{*}\hat{u}_{ij} \quad 1 \leq i,\ j \leq N \tag{4.131}$$

$a > 0$, P is a 3×3 matrix, so total 9 parameters have to be estimated. The objective function has been prepared for training the network which is the summation of squared error of all cells:

$$J = \frac{1}{2}\sum_{i=1}^{N}\sum_{j=1}^{N}(\hat{u}_{ij}(P,t) - u_{ij}^{*}(t))^{2} \tag{4.132}$$

where $u_{ij}^{*}(t)$ is the desired value of cell (i,j) at time t. Stable gradients are computed according to the objective function represented by Eq. (4.132) are as:

$$\frac{\partial J}{\partial P(m,n)} = \sum_{i=1}^{N}\sum_{j=1}^{N}\frac{\partial \hat{u}_{ij}}{\partial P(m,n)}\tilde{u}_{ij} \quad m,n = 1,2,3 \tag{4.133}$$

where, $\tilde{u}_{ij} = \hat{u}_{ij} - u_{ij}^{*}$. Then the derivatives of template are calculated with respect to (m,n) element of template P. After computing the gradients of the objective function, update rules are employed to achieve desired parameters. The update rule for each element of template is:

$$\dot{P}(m,n) = -\gamma \frac{\partial J}{\partial P(m,n)} - \rho||\tilde{u}||P(m,n) \quad (4.134)$$

$\gamma > 0$ is the learning rate and $\rho > 0$ is the damping rate. Simulation has been done for the heat equation using the modified training methodology and compared to the analytic solution. A CNN trained by modified back propagation algorithm, is capable of adjusting the parameters to model the dynamic of a heat equation even with large changes in boundary conditions, without any knowledge of system equations.

Authors in [99] proposed a concept on CNN paradigm for ultra fast, potentially low cost, and high precision computing of stiff partial differential equation and ordinary differential equations with cellular neural network.

4.4.3.10 NAOP for CNN Template Calculation

The concept is based on a straight forward scheme called Non linear adaptive optimization (NAOP), which is used for a precise template calculation for solving any nonlinear ordinary differential equation through CNN processor. The NAOP is performed by a complex computing module which works on two inputs, the first input contains wave solutions of models that describes the dynamics of a CNN network model built from state control templates:

$$\frac{dx_i}{dt} = -x_i + \sum_{j=1}^{M} [\hat{A}_{ij}\, x_j + \hat{A}_{ij}\, x_j + B_{ij}\, u_j] + I_i \quad (4.135)$$

And the second input contains the linear or nonlinear differential equation, under investigation which can be written in a flowing set of ordinary differential equations:

$$\frac{d^2 y_i}{d\,t^2} = F\left(y_i, y_i^n, y_i^m, z_i, z_i^n, z_i^m, t\right) \quad (4.136)$$

$$\frac{d^2 z_j}{dt^2} = F\left(z_j, z_j^n, z_j^m, y_j, y_j^n, y_j^m, t\right) \quad (4.137)$$

When the convergence process of the training process is achieved, the output of the NAOP system will generate after some training steps. The main benefit of solving ODE and PDE using CNN is the offered flexibility through NAOP to extract the CNN parameters through which CNN can solve any type of ODE or PDE.

In [100] a CNN model has been developed for solving set of two PDEs describing water flow channels called Saint Venant equation.

4.4.3.11 CNN for Solving Saint Venant 1D Equation

The set of partial differential equation which describes the problem involves two equations: First is the preserve mass equation

$$\frac{\partial S(x,t)}{\partial t} + \frac{\partial Q(x,t)}{\partial x} = q \tag{4.138}$$

And the second equation is the preserve momentum equation

$$\frac{\partial Q(x,t)}{\partial t} + \frac{\partial \left[\frac{Q(x,t)}{S(x,t)}\right]}{\partial x} + gS(x,t)\frac{\partial h(x,t)}{\partial x} - gIS(x,t) + gJS(x,t) = k_q q \frac{Q(x,t)}{S(x,t)} \tag{4.139}$$

For solving the above equation using CNN, templates have to be designed by choosing the difference space of variables x with step Δx. Scalar energy function has been written for the function h and for the function Q, also the stability of CNN system is proved by discovering the state and output of each cell. Solving Saint Venant equation, following advantages are obtained: As in the theory of Taylor's expansion if we get more derivative terms the approximation will reach closer to the original equation and learning algorithms can be used to find better templates from original by choosing grid steps and circuit parameters.

4.5 Method of Finite Element Neural Networks

A major drawback of all the above approaches is that the network architecture is arbitrarily selected, and the performance of the neural networks depends on the data used in training and testing. If the test data is similar to the training one, the network can interpolate between them otherwise the network is forced to extrapolate and the performance degrades.

Hence the solution to the problem is to combine the power of numerical models with the computational speed of neural networks. So, Takeuchi and Kosugi [101] developed a finite element neural network formulation (FENN) to overcome these difficulties. The FENN can be used to solve forward problem and can also be used in an iterative algorithm to solve inverse problems. Finite element neural network method for solving the differential equation is given in Eq. (3.63).

Initially the finite element model can be converted into a parallel network form. Let us take an example of solving typical inverse problem arising in electromagnetic nondestructive evaluation (NDE), but the basic idea is applicable to other areas as well. NDE inverse problems can be formulated as the problem of finding material properties within the domain of problem. Since the domain is discretize in the FEM method by a large number of elements, the problem can be posed as one of finding the material properties in each of these elements. These properties are usually embedded in the differential operator L, or equivalently, in the global matrix K. Thus in order to be able to iteratively estimate these properties from the measurements, the material properties needs to be separated out from K. This separation is easier to achieve at the element matrix level. For nodes i and j in element e.

$$\begin{aligned} K_{ij}^e &= \int\limits_{\Omega^e} N_i^e L N_j^e \, d\Omega \\ &= \int N_i^e \alpha^e \bar{L} N_j^e \, d\Omega \\ &= \alpha^e S_{ij}^e \end{aligned} \tag{4.140}$$

where α^e is the parameter representing the material property in the element e and $\bar{L}$ represents the differential operator at the element level without α^e embedded in it. From Eq. (4.140), we get the functional

$$F\left(\tilde{\phi}\right) = \sum_{e=1}^{M} \left(\frac{1}{2} \phi^{e^t} \alpha^e S^e \phi^e - \phi^{e^t} b^e \right) \tag{4.141}$$

If we define,

$$K_{ij} = \sum \alpha^e w_{ij}^e \tag{4.142}$$

where,

$$w_{ij}^e = \begin{cases} S_{ij}^e, & i,j \in e \\ 0, & else \end{cases} \tag{4.143}$$

$$\begin{aligned} 0 = \frac{\partial F}{\partial \Phi_i} &= \sum_{j=1}^{N} K_{ij} \Phi_j - b_i \\ &= \sum_{j=1}^{N} \left(\sum_{e=1}^{M} \alpha^e w_{ij}^e \right) \Phi_j - b_i, \quad i = 1, 2, \ldots, N \end{aligned} \tag{4.144}$$

Equation (4.144) expresses the functional explicitly in terms of α^e. This can be easily converted into a parallel network form and neural network comprises an input, output and hidden layer. In the general case with M elements and N nodes in the FEM mesh, the input layer with M network inputs takes the α values in each element as input. The hidden layer has N^2 neurons arranged in N groups of N neurons, corresponding to the N^2 members of the global matrix K. The output of each hidden layer neurons is the corresponding row vector of K. The weights from the input to the hidden layer are set to the appropriate values of w_{ij}^e. The output of the hidden layer neurons are the elements K_{ij} of the global matrix as given in Eq. (4.143).

Each group of hidden neurons is connected to one output neuron by a set of weights Φ, with each element of Φ representing the nodal values Φ_j. The set of weights Φ between the first group of hidden neurons and the first output neuron are same as the set of weights between the second group of hidden neurons and the second output neuron. Each output neuron is also a summation unit followed by a linear activation function, and the output of each neuron is equal to b_i as

$$b_i = \sum_{j=1}^{N} K_{ij}\Phi_j = \sum_{j=1}^{N} \Phi_j \left(\sum_{j=1}^{M} \alpha^e w_{ij}^e \right) \tag{4.145}$$

where the second part of Eq. (4.145) is obtained by using Eq. (4.144).

4.5.1 Boundary Conditions in FENN

The elements of K^s and b^s do not depend on the material properties α. K^s and b^s need to be added appropriately to the global matrix K and the source vector b. Thus natural boundary conditions can be applied in the FENN as bias inputs to the hidden layer neurons that are a part of the boundary, and the corresponding output neurons. Dirichlet boundary conditions are applied by clamping the corresponding weights between the hidden layer and output layer neurons. These weights are referred to as the clamped weights, while the remaining weight will be referred to as the free weights. In Refs. [103–110] finite element neural network has been considered for various kind of differential equations.

Remark 5 The FENN architecture can be derived without consideration of dimensionality of the problem at hand so we can use FENN for 1D, 2D, 3D, or higher dimensional problems. The FENN architecture has a weight structure that allows both the forward and inverse problems to be solved using simple gradient based algorithms. The major advantage of the FENN is that it represents the Finite element model in a parallel form, enabling parallel implementation in either hardware or software; computing gradient in the FENN is very simple and for solving inverse problems is that it avoids inverting the global matrix in each iteration. The FENN also does not require any training, since most of its weights can be computed in advance and stored. It also reduces the computational effort associated with the network.

4.6 Method of Wavelet Neural Networks

Wavelet neural networks are a new class of neural networks with unique capabilities in system identification and classification which was proposed as an alternative to the feed forward neural networks for approximating arbitrary non linear functions. It has become a popular tool for non linear approximation due to its properties. It not only has the properties of self organized, self learning and strong error tolerance of neural network but has properties of finite support and self similarity of wavelets. In [111] wavelet neural network method is used for solving steady convection dominated diffusion problem. In back propagation algorithm sigmoid functions are used to approximate the non linearity while in wavelet neural network non linearity is approximated by superposition of a series of wavelet functions. A wavelet transform V with respect to the function $f(x)$ can be expressed as:

$$V_f = |l|^{1/2} \int_{-\infty}^{+\infty} f(x)\phi\left(\frac{x-b}{a}\right)dx = \langle f(x), \phi_{l,m}(x)\rangle \tag{4.146}$$

where l and m are the dilation and translation factors. If l and m are discrete numbers then the transform is known as the discrete wavelet transform. The wavelet series expansion of the function can be expressed in the following given form:

$$f(x) = \sum_i \sum_j \theta_{ij}\phi_{ij}(x) \tag{4.147}$$

where

$$\theta_{ij} = \int_{-\infty}^{+\infty} f(x)\phi_{ij}(x)dx$$

when function ϕ is taken as the activation function of network it is called as the wavelet neural network. A three layered neural network can be constructed using the above function represented by Eq. (4.147) and assuming the number of neurons of input, hidden and output layer are respectively n, N and m, the input and output of the wavelet neural network in each layer can be given by

$$I_j = \frac{\sum_{i=1}^{n} w_{ji}x_i - b_j}{a_j}, O_j = \phi(I_j), \tag{4.148}$$

$$y_k = \sum_{j=1}^{N} w_{jk}O_j \quad \text{for } j = 1, 2, \ldots, N, \quad k = 1, 2, \ldots, m \tag{4.149}$$

The Dirichlet boundary value problem associated with steady convection diffusion transport is defined by the following equation:

$$a \cdot \nabla u - \nabla \cdot (v\nabla u) = k \text{ in } \Omega \tag{4.150}$$

$$u = \bar{u} \text{ on } \Gamma_D \tag{4.151}$$

In Eqs. (4.150) and (4.151) u is scalar unknown quantity, $a(x)$ is the convection velocity, $v > 0$ is coefficient of diffusion and $k(x)$ is the volumetric source term. In wavelet neural network method, consider the variables x of an unknown function g as the input of the WNN, and the derivatives with the highest order of the unknown function as the output of the WNN. The objective function for minimization problem can be constructed as:

$$E(x) = |a \cdot \nabla u - \nabla \cdot (v\nabla u) - k| \tag{4.152}$$

Integration process is applied for the other lower order derivatives of Eq. (4.152) and unknown function with respect to the variable x_i, and the integration constants that generated in the integration are evaluated by the boundary points. For solving the steady convection diffusion transport problem a fourth order scale function of spline wavelets has been chosen as activation function to test WNN which is:

$$\phi_4(x) = \frac{1}{6}\begin{cases} 0 & x \leq 0 \\ x^3 & x \in [0,1] \\ 4 - 12x + 12x^2 - 3x^3 & x \in [1,2] \\ -44 + 60x - 24x^2 + 3x^3 & x \in [2,3] \\ 64 - 48x + 12x^2 - x^3 & x \in [3,4] \\ 0 & x \geq 4 \end{cases} \tag{4.153}$$

Fourth order scale function is symmetric about $x = 2$ and compactly supported in the range $[0, 4]$. Numerical simulation has been done and particle swarm optimization technique is used to minimize the error quantity and the results obtained are closed to the exact solution for convection dominated diffusion problem.

Remark 6 The advantage of the wavelet neural network method is that, once the WNN is trained and its parameters are stored, it allows instantaneous evaluation of the solution at any desired point in the domain with spending negligible computing time. It can eliminate the singularly perturbed phenomenon in the equation and its precision is also high in learning process and prediction process. Work is in progress for solving differential equations using finite element and wavelet neural network.

4.7 Some Workout Examples

In this section we illustrate workout examples on some of the methods discussed above in this chapter:

Example 4.7.1 Let us consider a simple two point boundary value problem arising in the position of falling object as:

$$\frac{d^2y}{dt^2} + \frac{c}{m}\left(\frac{dy}{dt}\right) - g = 0 \tag{4.156}$$

where $c = a$ drag coefficient $= 12\,\text{kg/s}$, $m = 80$ kg, and $g =$ acceleration due to gravity $= 9.82\,\text{m/s}^2$ with the following boundary conditions: $y(0) = 0$, $y(12) = 600$.

Solution To obtain a neural network solution for the Eq. (4.156) along with the boundary conditions following steps are required:

Step 1: First we construct a trial solution of the neural network for Eq. (4.156) of the following form:

$$y_T(t,p) = A(t) + F(t, N(t,p))$$

where first term satisfies initial/boundary value problem and second term represents feed forward neural network with input vector x and p is the adjustable weight parameters. Hence, we propose a trial solution for Eq. (4.156) as:

$$y_T(t,p) = 50\,t + t(12 - t)\,N(t,p) \tag{4.157}$$

which satisfies the boundary conditions as:

$$y_T(0,p) = 50 \times 0 + 0(12 - 0)N(0,p) = 0$$

and

$$y_T(12,p) = 50 \times 12 + 12(12 - 12)N(12,p) = 600$$

Step 2: Since y_T is an approximate solution to Eq. (4.156) for optimized values of parameters p. Thus the problem of finding an approximate solution to Eq. (4.156) over some collocation points in the domain [0, 12] is equivalent to calculate the functional $y_T(t,p)$ that satisfies the constrained optimization problem. If we consider the trial solution of the following form given in Eq. (4.157) the problem is converted into an unconstrained optimization problem and the error quantity to be minimize can be given by the following equation:

$$E(t) = \sum_i \left\{ \frac{d^2 y_T(t_i,p)}{dt} - f\left(x_i, \frac{dy_T(t_i,p)}{dt}\right) \right\}^2 \tag{4.158}$$

where,

$$\frac{dy_T(t_i,p)}{dt} = (12-2t)N(t,p) + (12t-t^2)N'(t,p)$$

and,

$$\frac{d^2 y_T(t_i,p)}{dt^2} = -2N(t,p) + 2(12-2t)N'(t,p) + (12t-t^2)N''(t,p)$$

Step 3: Set up the network with randomly generated vector $x \in [0, 12]$ and $u_i, v_i, w_i \in [-0.5, 0.5]$ for $i = 1, 2, \ldots, h$ together with ε an error limit, where h is number of neurons in the hidden layer. For network parameter updation we compute derivative of neural network with respect to input as well as for parameters of network and train the neural network for optimized value of parameters.

Step 4: Once the network is trained set up the network with optimized network parameters and compute $y_T(t,p)$ from Eq. (4.157).

The neural network constructed for Eq. (4.156) is trained using a grid of almost 13 equidistant points and mean sum squared error is reduced to a minimum of 10^{-10}. So the estimated solution of Eq. (4.156) using neural network is given in Table 4.1.

Example 4.7.2 As the example of partial differential equations we consider the wave equation arising in non controlled longitudinal vibration of rod as:

$$\frac{\partial^2 y}{\partial t^2} - \alpha^2 \frac{\partial^2 y}{\partial x^2} = 0 \tag{4.159}$$

together with the following initial and boundary conditions with $t \in [0, 1]$ and $x \in [0, 1]$:

$$(x,0) = \sin(\pi x), \quad \frac{\partial y(x,0)}{\partial t} = 0, \quad 0 \le x \le 1$$

$$y(0,t) = y(1,t) = 0$$

Table 4.1 Neural network solution for Example 4.7.1

t	y	$\frac{dy}{dt}$
0.000	0.00108	32.0135
1.000	32.6934	36.7844
2.000	76.8842	40.3976
3.000	116.7498	44.5563
4.000	170.2756	46.5129
5.000	214.8462	48.5245
6.000	261.1654	51.7344
7.000	324.6534	54.0216
8.000	373.1183	55.6643
9.000	425.8863	56.2267
10.000	493.5689	58.2263
11.000	547.7762	59.1378
12.000	600.000	59.8624

Solution Following steps are required for solving the Eq. (4.159) using neural network:

Step 1: Construct a trial solution of neural network in the following form:

$$y_T(x,t,p) = A(x,t) + x(1-x)t(1-t)\,[N(x,t,p)]$$

Thus, assuming $\alpha = 1$ the trial solution of neural network can be written as:

$$y_T(x,t,p) = (1-t^2)\sin(\pi x) + x(1-x)t^2[N(x,t,p)] \tag{4.160}$$

which satisfies the boundary conditions as:

$$\begin{aligned}
y_T(x,0,p) &= (1-0^2)\sin(\pi x) + x(1-x)0^2[N(x,t,p)] = \sin(\pi x)\\
y_T(x,t,p) &= (1-t^2)\sin(\pi x) + x(1-x)t^2[N(x,t,p)]\\
y_T(0,t,p) &= (1-t^2)\sin(\pi\cdot 0) + 0(1-0)t^2[N(x,t,p)] = 0\\
y_T(1,t,p) &= (1-t^2)\sin(\pi\cdot 1) + 1(1-1)t^2[N(x,t,p)]
\end{aligned}$$

and

$$\frac{\partial y_T(x,0,p)}{\partial t} = 0$$

Step 2: Now the error function that has to be minimize can be given by the following equation:

$$E(\mathrm{p}) = \sum_i \left\{ \frac{\partial^2 y_T(x_i, t_i)}{\partial t^2} + \frac{\partial^2 y_T(x_i, t_i)}{\partial x^2} - f(x_i, t_i) \right\}^2 \tag{4.161}$$

Step 3: A neural network with one input layer, a hidden layer with h number of neurons together with an output layer is constructed in which weights and initial values are random parameters. A neural network is trained to optimize the network parameters.

Step 4: Once the network is trained the solution of the differential equation is obtained from Eq. (4.160) with optimized network parameters. The solution of Eq. (4.159) is given in the following Table 4.2.

Example 4.7.3 In this example we are considering the basic equation of beam column theory, linking the displacement of the centre line $u(x)$ to the axial compressive load F and the lateral load $l(x)$ [85] i.e.

$$EI\frac{d^4u}{dx^4} + F\frac{d^2u}{dx^2} = l. \tag{4.162}$$

together with the boundary conditions

(i) $$u(0) = u'(0) = u(k) = u'(k) = 0. \tag{4.163}$$

(ii) $$u(0) = u''(0) = u(k) = u''(k) = 0. \tag{4.164}$$

(iii) $$u(0) = u'(0) = u(k) = u''(k) = 0. \tag{4.165}$$

For the first case given by Eq. (4.163) trial solution can be given as:

$$u_T(x, K) = (x^4 + x^2k^2 - 2x^3k)N(x, K). \tag{4.166}$$

which satisfies the boundary conditions given in Eq. (4.6). So, the error is to be minimized in the following form:

Table 4.2 Neural network solution for Example 4.7.2

x	t	y
0.0000	0.0000	0.00000
0.5000	0.1000	1.00010
0.5000	0.4000	0.30869
0.4000	0.6000	−0.29943
0.8000	1.0000	−0.951056
1.0000	0.3000	0.071863

$$E(\bar{K}) = \left\{u_T^{(iv)}(x_i, K) - f\left(x_i, u'_T(x_i, K), u''_T(x_i, K), u'''_T(x_i, K)\right)\right\}. \tag{4.167}$$

For second case given by Eq. (4.164) we propose a trial solution of a beam column hinged at both ends is of the form:

$$u_T(x, K) = \left(\frac{16}{5}k^{-4}x^4 - \frac{32}{5}k^{-3}x^3 + \frac{16}{5}k^{-1}x\right)\left(\frac{N'_0 - N'_k}{2k}x^2 - N'_0 x + N\right). \tag{4.168}$$

where,

$$N'_0 = \left.\frac{dN}{dx}\right|_{x=0}, \quad \text{and} \quad N'_k = \left.\frac{dN}{dx}\right|_{x=k},$$

Trial solution given in Eq. (4.168) satisfies the boundary conditions given in Eq. (4.164) as:

$$\begin{aligned} u'_T(x, K) = &\left(\frac{64}{5}k^{-4}x^3 - \frac{96}{5}k^{-3}x^3 + \frac{16}{5}k^{-1}\right)\left(\frac{N'_0 - N'_k}{2k}x^2 - N'_0 x + N\right) + \cdots \\ &\left(\frac{16}{5}k^{-4}x^4 - \frac{32}{5}k^{-3}x^3 + \frac{16}{5}k^{-1}x\right)\left(\frac{N'_0 - N'_k}{2}x - N'_0 + N'\right), \end{aligned} \tag{4.169}$$

And the trial solution for Eq. (4.165) can be given as:

$$u_T(x, K) = \sin\left(\frac{2\pi x}{k}\right)\left(\frac{N_0 - N_k}{2k}x^2 - N'_0 x + xN'\right) \tag{4.170}$$

Numerical simulation has been done for case 1 and the maximum absolute error calculated in the deflection function of a beam column fixed at both the ends are presented in Table 4.3.

Table 4.3 Error in deflection of beam column fixed at both end Eq. (4.163)

Load (F)	Maximum absolute error in deflection of the beam column				
	l = 0.05	l = 0.10	l = 0.15	l = 0.20	l = 0.25
0	1.0832×10^{-9}	4.8266×10^{-12}	2.8631×10^{-10}	1.3275×10^{-10}	9.8276×10^{-10}
200	3.7170×10^{-6}	3.6832×10^{-6}	5.3678×10^{-6}	1.1495×10^{-5}	4.8440×10^{-6}
400	2.7530×10^{-5}	5.1240×10^{-7}	4.3627×10^{-5}	2.7713×10^{-5}	5.6425×10^{-6}
600	5.4860×10^{-5}	4.0406×10^{-6}	5.837×10^{-5}	1.2222×10^{-5}	3.1417×10^{-5}
800	3.4310×10^{-5}	4.2197×10^{-5}	4.6617×10^{-5}	7.3132×10^{-5}	1.6669×10^{-5}
1,000	3.3950×10^{-4}	3.7631×10^{-4}	5.9272×10^{-4}	8.8689×10^{-4}	7.0911×10^{-4}

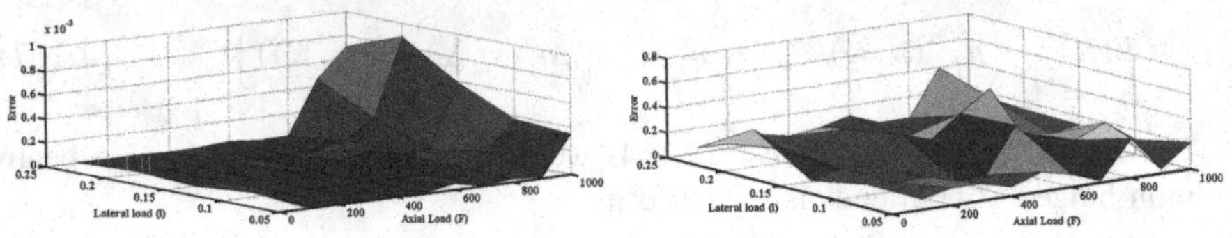

Fig. 4.4 Maximum absolute error and relative error in deflection of beam column fixed at both ends Eq. (4.163)

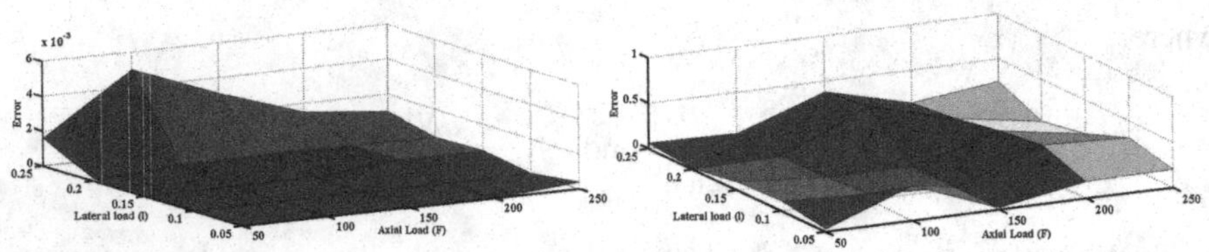

Fig. 4.5 Maximum absolute and relative error in the deflection of beam column hinged at both ends Eq. (4.164)

Table 4.4 Error in deflection of beam column fixed at both end Eq. (4.164)

Load	Maximum absolute error in deflection of the beam column				
(F)	1 = 0.05	1 = 0.10	1 = 0.15	1 = 0.20	1 = 0.25
0	1.1512×10^{-10}	6.2567×10^{-7}	3.5400×10^{-5}	5.845×10^{-8}	1.5320×10^{-9}
50	3.9100×10^{-6}	2.8580×10^{-5}	1.4180×10^{-5}	6.5100×10^{-6}	4.0770×10^{-5}
100	6.6600×10^{-6}	2.4900×10^{-5}	1.4180×10^{-5}	1.6310×10^{-6}	4.3010×10^{-5}
150	6.040×10^{-6}	2.4900×10^{-5}	1.419×10^{-5}	5.00×10^{-4}	5.00×10^{-4}
200	1.0353×10^{-4}	4.00×10^{-4}	5.00×10^{-4}	4.00×10^{-4}	4.00×10^{-4}
250	1.60×10^{-3}	5.0×10^{-3}	3.10×10^{-3}	1.5×10^{-3}	1.02×10^{-2}

The maximum absolute error and relative error calculated in the deflection of beam column fixed at both ends respectively as given by Figs. 4.4 and 4.5.

For the second case the calculated maximum absolute error is given in Table 4.4.

For the third case described in Eq. (4.165) the maximum absolute error is tabulated in table.

Example 4.7.8 Consider the reaction diffusion Eq. (4.171) mentioned in Ref. [112].

$$y'' + \lambda \exp\left[\frac{y}{(1+\alpha y)}\right] = 0, \quad t \in (0,1) \tag{4.171}$$

with the boundary conditions $y(0) = y(1) = 0$. The trial solution of Eq. (4.171) using neural network can be written as:

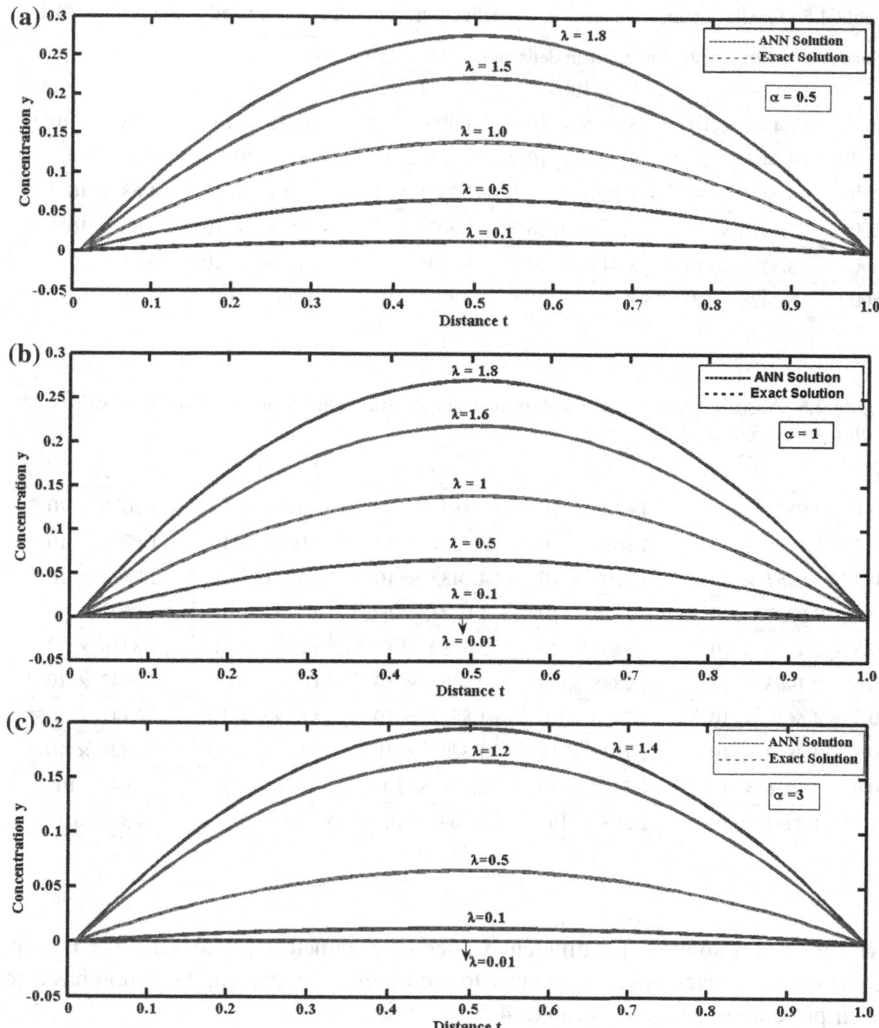

Fig. 4.6 Influence of λ on concentration $y(t)$ for $\alpha = 0.5$, 1 and 3 in Reaction Diffusion Equation using ANN

$$y_T = t(t-1)N(\bar{t}, \bar{p}) \tag{4.172}$$

which satisfies the desired boundary condition at $t = 0$ and 1. The derivatives are then calculated with respect to input vector and weight parameters to minimize the error quantity. We have considered three layered neural network with $h = 10$ number of hidden nodes and $N = 100$ (training points) to minimize the error term whose initial weights are chosen randomly. The ANN solution has been compared

Table 4.5 Maximum absolute error in the deflection of beam column fixed at the end Eq. (4.165)

Load (F)	Maximum absolute error in deflection of the beam column				
	1 = 0.05	1 = 0.10	1 = 0.15	1 = 0.20	1 = 0.25
0	1.4328×10^{-16}	5.8078×10^{-10}	4.6800×10^{-9}	5.675×10^{-12}	2.9603×10^{-10}
100	8.5466×10^{-4}	6.2320×10^{-4}	1.465×10^{-4}	1.7×10^{-3}	2.8×10^{-3}
200	5.2022×10^{-5}	1.1201×10^{-4}	3.2351×10^{-4}	5.3642×10^{-4}	4.63×10^{-4}
300	7.3016×10^{-5}	4.5322×10^{-5}	5.597×10^{-4}	4.5440×10^{-4}	4.82×10^{-4}
400	3.5087×10^{-5}	8.3430×10^{-5}	1.5545×10^{-4}	4.3966×10^{-4}	4.51×10^{-4}
500	6.3423×10^{-4}	1.8×10^{-3}	1.80×10^{-3}	1.06×10^{-3}	1.81×10^{-4}

Table 4.6 Absolute error in the solution of reaction diffusion equation for different values of λ with constant $\alpha = 3$

t	$\lambda = 0.01$	$\lambda = 0.1$	$\lambda = 0.5$	$\lambda = 1.2$	$\lambda = 1.4$
0.1	4.96535×10^{-6}	4.6302×10^{-5}	2.485×10^{-5}	1.2305×10^{-5}	1.1076×10^{-5}
0.2	3.1286×10^{-5}	4.8614×10^{-5}	3.4271×10^{-5}	1.0695×10^{-5}	1.8264×10^{-4}
0.3	3.0857×10^{-5}	3.8208×10^{-4}	4.9435×10^{-5}	1.1821×10^{-5}	2.0632×10^{-5}
0.4	3.9237×10^{-5}	3.4276×10^{-5}	1.0221×10^{-4}	2.3835×10^{-5}	2.8692×10^{-5}
0.5	2.4738×10^{-5}	4.3467×10^{-5}	2.3980×10^{-5}	1.2873×10^{-4}	1.3216×10^{-4}
0.6	2.1983×10^{-5}	1.2838×10^{-4}	2.0667×10^{-5}	1.76×10^{-4}	2.8642×10^{-4}
0.7	4.6650×10^{-5}	1.2690×10^{-5}	1.8532×10^{-4}	1.9243×10^{-4}	2.7147×10^{-4}
0.8	4.3852×10^{-5}	1.1741×10^{-5}	1.8391×10^{-4}	3.8216×10^{-5}	5.8838×10^{-4}
0.9	2.7036×10^{-4}	2.5472×10^{-4}	2.333×10^{-4}	6.3384×10^{-4}	2.864×10^{-5}
1.0	1.0831×10^{-4}	2.865×10^{-4}	1.269×10^{-4}	1.73×10^{-3}	2.634×10^{-4}

with the exact solution for different values of parameters α and λ for the present equations. Influence of one parameter to the other parameters and solution has also been presented in Fig. 4.6 (Tables 4.5 and 4.6).

Conclusion

Differential equations plays major role in applications of sciences and engineering. It arises in wide variety of engineering applications for e.g. electromagnetic theory, signal processing, computational fluid dynamics, etc. These equations can be typically solved using either analytical or numerical methods. Since many of the differential equations arising in real life application cannot be solved analytically or we can say that their analytical solution does not exist. For such type of problems certain numerical methods exists in the literature. In this book, our main focus is to present an emerging meshless method based on the concept of neural networks for solving differential equations or boundary value problems of type ODE's as well as PDE's. Here in this book, we have started with the fundamental concept of differential equation, some real life applications where the problem is arising and explanation of some existing numerical methods for their solution. We have also presented some basic concept of neural network that is required for the study and history of neural networks. Different neural network methods based on multilayer perceptron, radial basis functions, multiquadric functions and finite element etc. are then presented for solving differential equations. It has been pointed out that the employment of neural network architecture adds many attractive features towards the problem compared to the other existing methods in the literature. Preparation of input data, robustness of methods and the high accuracy of the solutions made these methods highly acceptable. The main advantage of the proposed approach is that once the network is trained, it allows evaluation of the solution at any desired number of points instantaneously with spending negligible computing time.

Moreover, different hybrid approaches are also available and the work is in progress to use better optimization algorithms. People are also working in the combination of neural networks to other existing methods to propose a new method for construction of a better trail solution for all kind of boundary value problems. Such a collection could not be exhaustive; indeed, we can hope to give only an indication of what is possible.

N. Yadav et al., *An Introduction to Neural Network Methods for Differential Equations*, SpringerBriefs in Computational Intelligence,
DOI 10.1007/978-94-017-9816-7

Appendix

Matlab Pseudo Code for the Solution of Differential Equation Using MLP Neural Network

```
1. **************** Training *****************************************************

function [G yT u v w] = DE3(u,v,w,xk,eta,F,g)
h = length(w);
[yT dyT d2yT sig sig1 sig2 sig3 sig4] = DE1(u,v,w,xk);

**************************** Preparation****************************************
G=0;
G = (d2yT + F*dyT - g);
for j=1:h
   dNdp      =       sig (j,:);
   d2Ndxdp   =    w(j)  *sig1(j,:);
   d3Ndx2dp  =    w(j)^2*sig2(j,:);
   d4Ndx3dp  =    w(j)^3*sig3(j,:);
   d5Ndx4dp  =    w(j)^4*sig4(j,:);
   dGdv      =    -2.*dNdp + 2.*(12-2.*xk).*d2Ndxdp + (12.*xk-xk.^2).*d3Ndx2dp +...
                   F.*((12-2.*xk).*dNdp + (12*xk-xk.^2).*d2Ndxdp)- g;


   dNdp      =     v(j)        *sig1(j,:);
   d2Ndxdp   =    v(j)  *w(j)  *sig2(j,:);
   d3Ndx2dp  =    v(j)  *w(j)^2*sig3(j,:);
   d4Ndx3dp  =    v(j)  *w(j)^3*sig3(j,:);
   d5Ndx4dp  =    v(j)  *w(j)^4*sig4(j,:);
   dGdu       =   -2.*dNdp + 2.*(12-2.*xk).*d2Ndxdp + (12.*xk-xk.^2).*d3Ndx2dp +...
                   F.*((12-2.*xk).*dNdp + (12*xk-xk.^2).*d2Ndxdp)- g;


   dNdp      =     v(j)*xk      .*sig1(j,:);
   d2Ndxdp   =     v(j)*xk*w(j)  .*sig2(j,:); +  v(j)    * sig1(j,:);
   d3Ndx2dp  =     v(j)*xk*w(j)^2.*sig3(j,:) + 2*v(j)*w(j)*sig2(j,:);
   d4Ndx3dp  =     v(j)*xk*w(j)^3.*sig3(j,:) + 3*v(j)*w(j)^2*sig3(j,:);
   d5Ndx4dp  =     v(j)*xk*w(j)^4.*sig4(j,:)+ 4*v(j)*w(j)^3*sig4(j,:);
   dGdw      =     -2.*dNdp + 2.*(12-2.*xk).*d2Ndxdp + (12.*xk-xk.^2).*d3Ndx2dp +...
                    F.*((12-2.*xk).*dNdp + (12*xk-xk.^2).*d2Ndxdp)- g;
```


N. Yadav et al., *An Introduction to Neural Network Methods for Differential Equations*, SpringerBriefs in Computational Intelligence,
DOI 10.1007/978-94-017-9816-7

```
****************************Updation****************************************

  v(j)        =        v(j) - eta*sum(2*G.*dGdv);
   u(j)       =        u(j) - eta*sum(2*G.*dGdu);
   w(j)       =        w(j) - eta*sum(2*G.*dGdw);
end
```

*************End*****************

References

1. H. Lee, I. Kang, Neural algorithms for solving differential equations. J. Comput. Phys. **91**, 110–117 (1990)
2. L. Wang, J.M. Mendel, Structured trainable networks for matrix algebra. IEEE Int. Jt. Conf. Neural Netw. **2**, 125–128 (1990)
3. D. Kincaid, W. Cheney, in *Numerical Analysis Mathematics of Scientific Computing*, 3rd edn. (American Mathematical Society, Providence, 2010)
4. A.J. Meade Jr., A.A. Fernandez, The numerical solution of linear ordinary differential equations by feedforward neural networks. Math. Comput. Model. **19**, 1–25 (1994)
5. A.J. Meade Jr., A.A. Fernandez, Solution of nonlinear ordinary differential equations by feedforward neural networks. Math. Comput. Model. **20**(9), 19–44 (1994)
6. M.E. Davis, *Numerical Methods and Modeling for Chemical Engineers* (Wiley, New York, 1984)
7. S. Haykin, *Neural Networks: A Comprehensive Foundation* (Pearson Education, Singapore, 2002)
8. J.M. Zurada, *Introduction to Artificial Neural Systems* (Jaico Publishing House, St. Paul, 2001)
9. R.H. Nielsen, *Neurocomputing* (Addison-Wesley Publishing Company, USA, 1990)
10. W.S. McCulloch, W. Pitts, A logical Calculus of the ideas immanent in nervous activity. Bull. Math. Biol. **5**, 115–133 (1943)
11. J.V. Neumann, *The General and Logical Theory of Automata* (Wiley, New York, 1951)
12. J.V. Neumann, Probabilistic logics and the synthesis of reliable organisms from unreliable components, in *Automata Studies* (Princeton University Press, Princeton, 1956), pp. 43–98
13. D.O. Hebb, *The Organization of Behaviour: A Neuropsychological Theory* (Wiley, New York, 1949)
14. F. Rosenblatt, *Principles of Neurodynamics* (Spartan Books, Washington, 1961)
15. M. Minsky, S. Papert, *Perceptrons* (MIT Press, Cambridge, 1969)
16. S. Amari, A theory of adaptive pattern classifiers. IEEE Trans. Electron. Comput. **16**(3), 299–307 (1967)
17. K. Fukushima, Visual feature extraction by multilayered networks of analog threshold elements. IEEE Trans. Syst. Sci. Cyber **5**(4), 322–333 (1969)
18. S. Grossberg, Embedding fields: a theory of learning with physiological implications. J. Math. Psychol. **6**, 209–239 (1969)
19. A.H. Klopf, E. Gose, An evolutionary pattern recognition network. IEEE Trans. Syst. Sci. Cyber **53**, 247–250 (1969)
20. J.J. Hopfield, Neural Networks and physical systems with emergent collective computational abilities. Proc. Natl Acad. Sci. **79**, 2254–2258 (1982)
21. J.J. Hopfield, Neurons with graded response have collective computational properties like those of two state neurons. Proc. Natl. Acad. Sci. **81**, 3088–3092 (1984)

N. Yadav et al., *An Introduction to Neural Network Methods for Differential Equations*, SpringerBriefs in Computational Intelligence,
DOI 10.1007/978-94-017-9816-7

22. D.E. Rumelhart, J.L. McClelland, *Parallel Distributed Processing: Explorations in the Microstructure of Cognition, I and II* (MIT Press, Cambridge, 1986)
23. M. Mahajan, R. Tiwari, *Introduction to Soft Computing* (Acne Learning Private Limited, New Delhi, 2010)
24. S. Pal, *Numerical Methods: Priniciples, Analyses and Algorithms* (Oxford University Press, Oxford, 2009)
25. L.O. Chua, L. Yang, Cellular neural networks: theory. IEEE Trans. Circuits Syst. **35**, 1257–1272 (1988)
26. Q. Zhang, A. Benveniste, Wavelet networks. IEEE Trans. Neural Netw. **3**, 889–898 (1992)
27. P.J. Werbos, Beyond regression: new tools for prediction and analysis in the behavioral Sciences, Ph.D. thesis, Harvard University, 1974
28. M. Reidmiller, H. Braun, A direct adaptive method for faster back propagation learning: the RPROP algorithm, in *Proceedings of the IEEE International Conference on Neural Networks* (1993), pp. 586–591
29. K.S. Mcfall, An artificial neural network method for solving boundary value problems with arbitrary irregular boundaries, Ph.D. thesis, Georgia Institute of Technology (2006)
30. V. Kecman, *Learning and Soft Computing* (The MIT Press, Cambridge, 2001)
31. D.J. Montana, L. Davis, Training feed forward neural networks using Genetic algorithms, in *Proceedings of the 11th International Joint Conference on Artificial Intelligence*, vol. 1 (1989), pp. 762–767
32. R.S. Sexton, J.N.D. Gupta, Comparative evaluation of genetic algorithm and back propagation for training neural networks. Inf. Sci. **129**, 45–59 (2000)
33. J.A. Khan, R.M.A. Zahoor, I.M. Qureshi, Swarm intelligence for the problems of non linear ordinary differential equations and its application to well known Wessinger's equation. Eur. J. Sci. Res. **34**, 514–525 (2009)
34. A. Yadav, K. Deep, A new disc based particle swarm optimization. Adv. Intell. Soft Comput. **130**, 23–30 (2012)
35. K. Hornik, M. Stinchcombe, H. White, Multilayer feedforward networks are universal approximators. Neural Netw. **2**(5), 359–366 (1989)
36. K. Hornik, M. Stinchcombe, H. White, Universal approximation of an unknown mapping and its derivatives using multilayer feedforward networks. Neural Netw. **3**, 551–560 (1990)
37. I.E. Lagaris, A.C. Likas, Artificial neural networks for solving ordinary and partial differential equations. IEEE Trans. Neural Netw. **9**, 987–1000 (1998)
38. S. He, K. Reif, R. Unbehauen, Multilayer networks for solving a class of partial differential equations. Neural Netw. **13**, 385–396 (2000)
39. I.E. Lagaris, A.C. Likas, D.G. Papageorgiou, Neural-network methods for boundary value problems with irregular boundaries. IEEE Trans. Neural Netw. **11**(5), 1041–1049 (2000)
40. L.P. Aarts, P.V. Veer, Neural network method for partial differential equations. Neural Process. Lett. **14**, 261–271 (2001)
41. N. Smaoui, S. Al-Enezi, Modeling the dynamics of non linear partial differential equations using neural networks. J. Comput. Appl. Math. **170**, 27–58 (2004)
42. A. Malek, R.S. Beidokhti, Numerical solution for high order differential equations using a hybrid neural network-optimization method. Appl. Math. Comput. **183**, 260–271 (2006)
43. J.A. Nelder, R. Mead, A simplex method for function minimization. Comput. J. **7**, 308–313 (1965)
44. Y. Shirvany, M. Hayati, R. Moradian, Numerical solution of the nonlinear Schrodinger equation by feedforward neural networks. Commun. Nonlinear Sci. Numer. Simul. **13**, 2132–2145 (2008)
45. R.S. Beidokhti, A. Malek, Solving initial-boundary value problems for systems of partial differential equations using neural networks and optimization techniques. J. Franklin Inst. **346**, 898–913 (2009)
46. I.G. Tsoulos, D. Gavrilis, E. Glavas, Solving differential equations with constructed neural networks. Neurocomputing **72**, 2385–2391 (2009)

47. I.G. Tsoulos, D. Gavrilis, E. Glavas, Neural network construction and training using grammatical evolution. Neurocomputing **72**, 269–277 (2008)
48. K.S. Mcfall, J.R. Mahan, Artificial neural network method for solution of boundary value problem with exact satisfaction of arbitrary boundary conditions. IEEE Trans. Neural Netw. **20**(8), 1221–1233 (2009)
49. A.G.L. Zagorchev, Acomparative study of transformation functions for non rigid image registration. IEEE Trans. Image Process. **15**(3), 529–538 (2006)
50. H. Alli, A. Ucar, Y. Demir, The solutions of vibration control problem using artificial neural networks. J. Franklin Inst. **340**, 307–325 (2003)
51. H. Saxen, F. Pettersson, Method for the selection of inputs and structure of feedforward neural networks. Comput. Chem. Eng. **30**, 1038–1045 (2006)
52. C. Filici, Error estimation in the neural network solution of ordinary differential equations. Neural Netw. **23**, 614–617 (2010)
53. P.E. Zadunaisky, On the estimation of errors propagated in the numerical integration of ordinary differential equations. Numer. Math. **27**, 21–39 (1976)
54. P.E. Zadunaisky, On the accuracy in the numerical solution of the N-body problem. Celest. Mech. **20**, 209–230 (1979)
55. V. Dua, An artificial neural network approximation based decomposition approach for parameter estimation of system of ordinary differential equations. Comput. Chem. Eng. **35**, 545–553 (2011)
56. N.K. Masmoudi, C. Rekik, M. Djemel, N. Derbel, Two coupled neural network based solution of the Hamilton-Jacobi-Bellman equation. Appl. Soft Comput. **11**, 2946–2963 (2011)
57. L. Jianyu, L. Siwei, Q. Yingjian, H. Yaping, Numerical Solution of differential equations by radial basis function neural networks. Proc. Int. Jt Conf. Neural Netw. **1**, 773–777 (2002)
58. J.E. Moody, C. Darken, Fast learning in networks of locally tuned processing units. Neural Comput. **1**(2), 281–294 (1989)
59. A. Esposito, M. Marinaro, D. Oricchio, S. Scarpetta, Approximation of continuous and discontinuous mappings by a growing neural RBF-based algorithm. Neural Netw. **13**, 651–665 (2000)
60. J. Park, I.W. Sandberg, Approximation and radial basis function networks. Neural Comput. **5**, 305–316 (1993)
61. R. Franke, Scattered data interpolation: tests of some methods. Math. Comput. **38**(157), 181–200 (1982)
62. N. Mai-Duy, T. Tran-Cong, Approximation of function and its derivatives using radial basis function networks. Neural Netw. **14**, 185–199 (2001)
63. N. Mai-Duy, T. Tran-Cong, Numerical solution of differential equations using multiquadric radial basis function networks. Neural Netw. **14**, 185–199 (2001)
64. T. Nguyen-Thien, T. Tran-Cong, Approximation of functions and their derivatives: a neural network implementation with applications. Appl. Math. Model. **23**, 687–704 (1999)
65. T.L. Lee, Back-propagation neural network for the prediction of the short-term storm surge in Taichung harbor, Taiwan. Eng. Appl. Artif. Intell. **21**, 63–72 (2008)
66. J. Rashidhinia, R. Mohammadi, R. Jalilian, Cubic spline method for two-point boundary value problems. IUST Int. J. Eng. Sci. **19**(5–2), 39–43 (2008)
67. K. Deng, Z. Xiong, Y. Huang, The Galerkin continuous finite element method for delay differential equation with a variable term. Appl. Math. Comput. **186**, 1488–1496 (2007)
68. M. Kumar, H.K. Mishra, P. Singh, A boundary value approach for singularly perturbed boundary value problems. Adv. Eng. Softw. **40**(4), 298–304 (2009)
69. N. Mai-Duy, T. Tran-Cong, Mesh free radial basis function network methods with domain decomposition for approximation of functions and numerical solution of Poisson's equations. Eng. Anal. Boundary Elem. **26**, 133–156 (2002)
70. L. Jianyu, L. Siwei, Q. Yingjian, H. Yaping, Numerical solution of elliptic partial differential equation by radial basis function neural networks. Neural Netw. **16**, 729–734 (2003)

71. E.J. Kansa, H. Power, G.E. Fasshauer, L. Ling, A volumetric integral radial basis function method for time dependent partial differential equations. I. formulation. Eng. Anal. Boundary Elem. **28**, 1191–1206 (2004)
72. H. Zou, J. Lei, C. Pan, Design of a new kind of RBF neural network based on differential reconstruction. Int. Jt. Conf. Neural Netw. Brain **1**, 456–460 (2005)
73. N. Mai-Duy, Solving high order ordinary differential equations with radial basis function networks. Int. J. Numer. Methods Eng. **62**, 824–852 (2005)
74. N. Mai-Duy, T. Tran-Cong, Solving biharmonic problems with scattered-point discretization using indirect radial basis function networks. Eng. Anal. Boundary Elem. **30**, 77–87 (2006)
75. A. Golbabai, S. Seifollahi, Radial basis function networks in the numerical solution of linear integro-differential equations. Appl. Math. Comput. **188**, 427–432 (2007)
76. A. Golbabai, M. Mammadov, S. Seifollahi, Solving a system of nonlinear integral equations by an RBF network. Comput. Math. Appl. **57**, 1651–1658 (2009)
77. A. Aminataei, M.M. Mazarei, Numerical solution of Poisson's equation using radial basis function networks on the polar coordinate. Comput. Math. Appl. **56**, 2887–2895 (2008)
78. H. Chen, L. Kong, W. Leng, Numerical solution of PDEs via integrated radial basis function networks with adaptive training algorithm. Appl. Soft Comput. **11**, 855–860 (2011)
79. S. Sarra, Integrated radial basis functions based differential quadrature method and its performance. Comput. Math. Appl. **43**, 1283–1296 (2002)
80. M. Kumar, N. Yadav, Multilayer perceptrons and radial basis function neural network methods for the solution of differential equations: A survey. Comput. Math. Appl. **62**, 3796–3811 (2011)
81. L.O. Chua, L. Yang, Cellular neural networks: theory. IEEE Trans. Circuits Syst. **35**, 1257–1272 (1988)
82. G. Manganaro, P. Arena, L. Fortuna, *Cellular neural networks: chaos, complexity and VLSI processing* (Springer, Berlin, 1999), pp. 44–45
83. J.C. Chedhou, K. Kyamakya, Solving stiff ordinary and partial differential equations using analog computing based on cellular neural networks. ISAST Trans. Comput. Intell. Syst. **1** (2), 38–46 (2009)
84. R. Brown, Generalizations of the Chua equations. IEEE Trans. Circuits Syst. I **40**, 878–884 (1993)
85. M. Kumar, N. Yadav, Buckling analysis of a beam column using multilayer perceptron neural network technique. J. Franklin Inst. **350**(10), 3188–3204 (2013)
86. C.A. Brebbia, J.C.F. Telles, L.C. Wrobel, *Boundary Element Techniques: Theory and Application In Engineering* (Springer, Berlin, 1984)
87. R.D. Cook, D.S. Malkus, M.E. Plesha, *Concepts and Applications of Finite Element Analysis* (Wiley, Toronto, 1989)
88. R.V. Dukkipati, *Applied Numerical Methods Using MATLAB* (New Age International Publisher, New Delhi, 2011)
89. M. Kumar, Y. Gupta, Methods for solving singular boundary value problems using splines: a survey. J. Appl. Math. Comput. **32**, 265–278 (2010)
90. T. Kozek, T. Roska, A double time scale CNN for solving two dimensional Navier-Stokes equation. Int. J. Circuit Theory Appl. **24**(1), 49–55 (1996)
91. D. Gobovic, M.E. Zaghloul, Analog cellular neural network with application to partial differential equations with variable mesh size. IEEE Int. Symp. Circuits Syst. **6**, 359–362 (1994)
92. T. Roska, L.O. Chua, T. Kozek, R. Tetzlaff, F. Puffer, Simulating non linear waves and partial differential equations via CNN-Part I: basic techniques. IEEE Trans. Circuits Syst. I Fundam. Theory Appl. **42**, 807–815 (1995)
93. T. Roska, L.O. Chua, T. Kozek, R. Tetzlaff, F. Puffer, K. Lotz, Simulating non linear waves and partial differential equations via CNN-Part II: typical examples. IEEE Trans. Circuits Syst. I Fundam. Theory Appl. **42**, 816–820 (1995)

94. F. Pufser, R. Tetzlafs, D. Wolf, A learning algorithm for cellular neural networks (CNN) solving nonlinear partial differential equations, in *Proceeding of International Symposium of Signals, Systems, and Electronics* (1995), pp. 501–504
95. A. Rasmussen, M.E. Zaghloul, CMOS analog implementation of cellular neural network to solve partial differential equations with a micro electromechanical thermal interface, in *Proceedings of the 40th Midwest Symposium on Circuits and Systems*, vol. 2 (1997), pp. 1326–1329
96. I. Krstic, B. Reljin, P. Kostic, Cellular neural network to model and solve direct non linear problems of steady state heat transfer, in *International Conference on EUROCON'2001, Trends in Communications*, vol. 2 (2001), pp. 420–423
97. S.T. Moon, B. Xia, R.G. Spencer, G. Han, E. Sanchez-Sinencio, VLSI implementation of a neural network for solving linear second order parabolic PDE, in 43*rd IEEE Midwest Symposium on Circuits and Systems* (2000), pp. 836–839
98. M.J. Aein, H.A. Talebi, Introducing a training methodology for cellular neural networks solving partial differential equations, in *Proceedings of International Joint Conference on Neural Networks* (2009), pp. 72–75
99. J.C. Chedjou, K. Kyamakya, U.A. Khan, M.A. Latif, Potential contribution of CNN-based solving of stiff ODEs & PDEs to enabling real-time computational engineering,in 12th *International Workshop on Cellular Nanoscale Networks and their Applications* (2010), pp. 1–6
100. V.D. Thai, P.T. Cat, Equivalence and stability of two layered cellular neural network solving saint venant 1D equation, in 11th *International Conference Control, Automation, Robotics and Vision* (2010), pp. 704–709
101. J. Takeuchi, Y. Kosugi, Neural network representation of the finite element method. Neural Netw. **7**(2), 389–395 (1994)
102. P. Ramuhalli, L. Udpa, S.S. Udpa, Finite element neural networks for solving differential equations. IEEE Trans. Neural Netw. **16**(6), 1381–1392 (2005)
103. A.I. Beltzer, T. Sato, Neural classification of finite elements. Comput. Struct. **81**, 2331–2335 (2003)
104. B.H.V. Topping, A.I. Khan, A. Bahreininejad, Parallel training of neural networks for finite element mesh decomposition. Comput. Struct. **63**(4), 693–707 (1997)
105. L. Manevitz, A. Bitar, D. Givoli, Neural network time series forecasting of finite-element mesh adaptation. Neurocomputing **63**, 447–463 (2005)
106. H. Jilani, A. Bahreininejad, M.T. Ahmadi, Adaptive finite element mesh triangulation using self-organizing neural networks. Adv. Eng. Softw. **40**, 1097–1103 (2009)
107. O. Arndt, T. Barth, B. Freisleben, M. Grauer, Approximating a finite element model by neural network prediction for facility optimization in groundwater engineering. Eur. J. Oper. Res. **166**, 769–781 (2005)
108. S. Koroglu, P. Sergeant, N. Umurkan, Comparison of analytical, finite element and neural network methods to study magnetic shielding. Simul. Model. Pract. Theory **18**, 206–216 (2010)
109. J. Denga, Z.Q. Yueb, L.G. Thamb, H.H. Zhuc, F. Huangshan, Pillar design by combining finite element methods, neural networks and reliability: a case study of the copper mine, China. Int. J. Rock Mech. Min. Sci. **40**, 585–599 (2003)
110. L. Ziemianski, Hybrid neural network finite element modeling of wave propagation in infinite domains. Comput. Struct. **81**, 1099–1109 (2003)
111. X. Li, J. Ouyang, Q. Li, J. Ren, Integration wavelet neural network for steady convection dominated diffusion problem, in *3rd International Conference on Information and Computing*, vol. 2 (2010), pp. 109–112
112. N. Yadav, A. Yadav, K. Deep, Artificial neural network technique for solution of nonlinear elliptic boundary value problems, in *Proceedings of Fourth International Conference on Soft Computing for Problem Solving, Advances in Intelligent Systems and Computing* vol. 335 (2015), pp. 113–121

Index

Note: Page numbers followed by "f" and "t" indicate figures and tables respectively

N. Yadav et al., *An Introduction to Neural Network Methods for Differential Equations*, SpringerBriefs in Computational Intelligence,
DOI 10.1007/978-94-017-9816-7

Printed in the United States
By Bookmasters